AIR UNIVERSITY

In Service to the Nation

Air Force Research Institute Strategic Concept for 2018–2023

John A. Shaud, PhD
General, USAF, Retired

Air University Press
Maxwell Air Force Base, Alabama

January 2009

Muir S. Fairchild Research Information Center Cataloging Data

Shaud, John A.
In service to the nation : Air Force Research Institute strategic concept for 2018–2023 / John A. Shaud.
p. ; cm.
Includes bibliographical references.
ISBN 978-1-58566-189-3
1. Military planning—United States. 2. United States—Air Force—Forecasting. 3. Strategic planning—United States—Military aspects. 4. United States—Air Force—Operational readiness. 5. Cyberterrorism—United States—Prevention. 6. Technological innovations—United States—Forecasting. I. Title.

358.41/684—dc22

Disclaimer

This study is a product of the Air Force Research Institute and represents an academic effort in response to a request from the Air University commander. The report's contents reflect the opinions of the authors and do not represent official Air Force views.

The Air Force Research Institute stood up within Air University, Air Education and Training Command, Maxwell AFB, Alabama, on 19 May 2008.

Comments or questions concerning this study or other research can be sent to

Director
Air Force Research Institute
155 N. Twining St., Bldg. 693
Maxwell AFB, AL 36112-6026

Contents

Foreword

In September 2007, Lt Gen Stephen R. Lorenz, then Air University commander, tasked the inactivated College of Aerospace Doctrine, Research and Education, and in turn, the newly formed Air Force Research Institute, to provide its perspective on what the United States Air Force (USAF) should look like in 10–15 years, roughly 2018–2023. Specifically, he directed the study should not be a detailed discussion of platforms or acquisitions, but instead a detailed discussion of the enduring attributes of the nation's air, space, and cyberspace forces in order to develop those constants as a framework for understanding how the USAF would serve the nation in the future.

The timeframe selected for this study establishes boundaries to good effect. It is not so near the current budget cycles that findings are constrained simply to linear extrapolations of today's programmatic details, but, at the same time, it is not so far in the future that today's global and national baselines and trends can be ignored.

There is much that is uncertain about the future. What is certain, however, is that the programs of today will have an impact upon the future. For example, the systems currently in design and those in the USAF inventory today will, by and large, be the systems available for use in warfare between 2018 and 2023. With informed decision making and bold leadership, the USAF will remain the world's preeminent air force, prepared to fly, fight, and win in service to the nation.

John A. Shaud, PhD
General, USAF, Retired
Director, Air Force Research Institute

Allen G. Peck
Lieutenant General, USAF
Commander, Air University

Executive Summary

On 12 September 2007, the Air University commander tasked the Air Force Research Institute to provide an "outside-the-beltway" look at what the United States Air Force (USAF) should be about in the future, specifically what the service should look like 10–15 years from now. This timeframe was far enough outside the Fiscal Year Defense Plan to avoid some current programmatic boundaries, but not so far out as to be immune to current trends. Additionally, constraining the "future" to 15 years limits the rather miraculous invention of extraordinary weapons found in some war games and other future studies—for example, rods from space—and similarly bounds the geo-political landscape.

The goal of this study is to identify the enduring attributes of our nation's air, space, and cyberspace force in context of major transitions. The tasking construct identified that the study should avoid a focus on hardware and resourcing, and implied the focus should be on roles, missions, and functions "such as the transition from the Cold War to Long War—or whatever it is to be called—era." Finally, the tasking identified the study's target audience as the presidential transition teams, with a delivery date "prior to the next election."

Strategy in the post–Cold War era remains fluid. Without a focus on a single foe, it is often complicated and debatable. This study attempts to neither reinvent the USAF nor protect the status quo at the expense of conventional wisdom. The intent is to understand the value of the service's contribution to national security and, where appropriate, offer considerations for change. This study is informed by an understanding of airpower history and a realization of current USAF systems and strategies. It is not meant to be all encompassing, but rather provides insight into the most pressing issues facing the USAF in the post–Cold War era. Binding the strategy together is a redefinition of Global Vigilance, Global Reach, and Global Power; not in a weapons systems–specific context, but rather as a framework through which the USAF serves the nation.

An underlying assumption is that while major combat operations, also known as conventional campaigns, are the most dangerous to America's national interest, the conduct of irregular warfare (IW) is the most likely. Accordingly, beyond recasting Global Vigilance, Reach, and Power, the study focuses on how the USAF can contribute to winning the current fight, while simultaneously maintaining the technological superiority necessary for prevailing in the future fight. In relation to the current fight, this study discusses winning IW, air mobility, air-ground integration and the planning cycle, intelligence reform, the transformation of airpower thought—including integration of unmanned aerial systems, the Total Force, and air base disaster contingency planning. In maintaining USAF technological superiority, the study focuses on acquisition reform, nuclear surety and deterrence, intercontinental ballistic missile (ICBM) and bomber modernization, space defense in depth, and operationally responsive space and cyberspace operations.

Chapter 1

The USAF Today

If you don't know where you are going, you might wind up someplace else.

—Yogi Berra

The USAF finds itself on unfamiliar ground. It has been continually engaged in combat operations for more than 17 years. It is attempting to recapitalize the oldest fleet in its history with little public support—the public has difficulty accepting that systems like the F-22 and F-35 are required in significant numbers when they perceive no global military threat. Meanwhile, USAF international commitments are growing despite fewer Airmen. Essentially, the USAF has simultaneously attempted to recapitalize, conduct offensive operations, and manage a drawdown, any one of which would fully consume a service. The confidence Congress has had in the USAF in the past is under attack, as evidenced by the most recent tanker controversy; cancellation of space systems; and a congressional initiative, not the USAF, insisting on a new generation bomber. Finally, loss of command and control over nuclear components and weapons, capped by the historical event of losing a chief and secretary at the same time, presents major challenges for the USAF.

Despite these challenges, the USAF remains the most capable and lethal air force on the globe. Daily, it employs systems that are recognized internationally as "world class." It has professional, highly motivated Airmen who are the most educated officer and noncommissioned officer (NCO) corps in the world, again envied by all others. The world comes to the USAF for airlift, while the public and militaries worldwide rely upon Air Force systems for navigation, communications, and critical economic and financial services. So, taken in context, what the USAF requires to meet today's challenges while preparing for future opportunities is decisive leadership that leverages continuously developing innovative solutions.

Change begins with introspection, just as an effective strategy begins with a critical analysis of the current environment. In numerous areas, the USAF has positioned itself properly to succeed in the 2018–23 timeframe. This study does not attempt to address these areas. Rather, in special areas of concern, this strategy examines the current environment and makes recommendations that will better prepare the USAF to meet the challenges of the next 10 to 15 years.

Some will argue that this study is too founded on the issues facing the USAF today and is not sufficiently focused on a strategic vision of the USAF's future. This argument ignores the reality that the USAF is facing problems today that, if not resolved in the near-term, will adversely impact the 2018–23 timeframe. Further, it ignores the fact that the programs the USAF will need in 2018–23 must originate in this decade to be implemented 10 to 15 years from now. Any forward–thinking study must be founded in the present to appreciate the issues and accordingly, address viable strategies for a possible future.

The USAF has long struggled with what is an appropriate strategy for the post–Cold War era, one that helps shape its unique capabilities to meet the needs of the nation. Every secretary and chief must fight a "current fight"—the current fight during their tenure—and prepare the service for a future fight. This study proposes such a strategy, one founded in today's realities, but one that prepares the USAF to successfully meet its near-term commitments, while providing a direction for future success.

In developing a strategy, this study proposes redefining the concepts of Global Vigilance, Global Reach, and Global Power; a redefinition necessary for a service attempting to ensure superiority in the three domains of air, space, and cyberspace. Further, this effort focuses on the means that will enable the USAF to be successful in the 2018–23 timeframe, while making recommendations on how the USAF can continue to innovate and maintain its technological superiority in preparation for prevailing against future adversaries.

Chapter 2

Framing the Question

The mission of the United States Air Force is to fly, fight and win . . . in air, space, and cyberspace.

—Air Force Mission Statement
Approved December 2005

Predicting the future is a difficult task. Often such studies reach too far, attempting to predict "a work in progress." For example, America is still waiting for the flying cars and robotic house servants depicted at the 1939 New York World's Fair. So, what will the USAF look like in 10 to 15 years? One senior leader believes it will look "about 80 percent of what it looks like today." A basis for this conclusion is to compare today with 10 to 15 years ago. Certainly, America has seen the explosion of new technologies, such as the Internet, cell phones, hybrid cars, and the B-2 and F-22 wings, but the mainstays of today's USAF systems were available during Operation Desert Storm.

So, what will the US and international settings look like during the 2018–23 timeframe that will dictate the operating environment for the USAF? First and foremost, the United States will continue to be engaged globally and will not have the occasion of reverting to the pre–World War II isolationist national security strategy. The Global War on Terror (GWOT)—possibly more precisely renamed at a later date—will continue and there will be no break from high operational and personnel tempo missions and expeditionary deployments. The continuation of the GWOT will require the USAF to organize itself to best meet the irregular warfare (IW) challenge. From a global perspective, no single nation will emerge as a near-peer competitor; regional actors will challenge US national interests, and, at a specific location and time, may emerge as a near-peer competitor.

A constrained Department of Defense (DOD) budget (see Appendix B) combined with virtually unchanged Title 10 requirements will dictate USAF actions in the 2018–23 timeframe. American support for major military programs will be uncertain. At the same time, the public will expect the USAF to sustain and enhance its competencies in air, space, and cyberspace. Future training must be consistent with preparing for not just the present war, but any future war. Finally, recapitalization of platforms and personnel under a constrained budget will require developing new acquisition paradigms as the USAF struggles to equip itself for critical warfighting missions across the global domains of air, space, and cyberspace. As a result, the USAF must re-evaluate systems and conclude that "right tech" and not "high tech" will often be the desired solution.

Globalization will continue and the European Union, China, India, Indonesia, and Russia will challenge the US market advantages. The economic importance of each will continue to grow globally, placing additional pressure on the dollar at home and abroad. The Chinese economy will increase at a rate exceeding the international average, becoming more volatile as the central government struggles with the pressures of a growing middle class and demands for individual freedoms. China's emerging economic might will drive increased military development, moving it closer to dominance among its Asian neighbors. Russia will awaken from a period of decline as the world's insatiable demand for hydrocarbon energy continues to transfer significant wealth to both Russia and the Organization of Petroleum Exporting Countries (OPEC) states. While a return to a strategic cold war is unlikely, Russia will flex its new-found wealth by attempting to exert its influence across its vast western and southern borders—driven by a century's long preoccupation with protecting itself *via* a regional zone of influence. However, the demographic concerns of a low birth rate will continue to influence Russia's foreign policy.

Due to massive human capital wealth, both China's and India's economies, and conversely their militaries, will grow at a rate exceeding the international average. Further, relations between India and Pakistan will become a greater concern—particularly in the nuclear arena. The Pakistani government will continue to waver between democracy, military rule, and radical extremism, contributing to continued regional instability.

Nations near the equator will play a greater international role. Brazil will emerge as the regional power in Latin America—driven by its vast natural resources and renewable energy research. The United States will find itself increasingly involved in Africa, working to preclude failed states from serving as breeding grounds for radical extremists. US foreign policy will continue to focus on promoting stability within the Middle East. Internal policies, radical Islam, and the growing cost of energy will keep US interests linked to the stability of the Saudi dynasty and the Gulf States.

Strategic resource competition will exacerbate regional tensions. Climate change; famine; disease; population growth and migration; availability of petroleum, water, and strategic minerals; religious extremism; and the volatile financial markets will overwhelm nations, causing many to cede sovereign authority to international bodies or face Balkanization. A growing global population imbalance will accelerate between North and South—as well as East and West—increasingly highlighting the distinction between the "haves" and "have–nots."

The technological revolution of the twentieth and early twenty-first centuries will continue, and accelerate. While the United States will not lose its preeminent technological lead, that lead will diminish. Nuclear, biological, and chemical proliferation, despite America's commitments to counter-proliferation, will accelerate as weapons of mass destruction become more readily available to both nation–states and non-state actors. Space will become increasingly cluttered. China, India, and Russia will further exploit the medium—complicating space control and assured access to space. The resulting increase in platforms and debris will place constraints on US systems not heretofore experienced. Cyberspace, the new frontier of the early twenty-first century, will mature as a medium, simultaneously creating opportunities and exposing significant economic and military vulnerabilities.

Chapter 3

The Current Fight

> *Have a plan for everything. A plan for practice, a plan for the game. A plan for being ahead, and a plan for being behind 20–0 at the half, with your quarterback hurt and the phones dead, with it raining cats and dogs and no rain gear because the equipment man left it at home.*
>
> —Coach Paul "Bear" Bryant

As a service, the USAF must plan for and acquire systems, people, and the enabling structures and processes necessary to deter, fight, and win conventional campaigns (also known as major contingency operations)—foundational for US sovereignty—and to play its part in winning the Long War. Should conventional campaign deterrence fail, for any reason, the USAF must be prepared, and ready, to engage as part of a team in winning that fight. But the probability of conventional campaigns, given our dedication to maintaining a deterrent edge, is less likely than the continuation of irregular styles of warfare characteristic of the Long War.

By maintaining a dominant posture for conventional campaigns, America, in essence, has dictated where it will not fight. This does not completely rule out the potential for irrational choices or choices borne of desperation that result in a short, intense conflict. In this timeframe, however, the probability is low for a state-on-state, force-on-force attrition war challenging US sovereignty or a significant realignment of national/regional power—assuming the USAF retains current readiness for conventional campaigns.

In transforming the USAF's structure for the future, it must be clear that in relation to conventional campaigns, the nation must not forfeit its current advantages. In 2003, the Iraqis were so confounded by USAF air supremacy that they buried front line fighters valued at hundreds of millions of dollars, including some fairly sophisticated electronic warfare assets. To deter conventional campaigns, this portrait of formidable American airpower must remain crystal clear for our adversaries.

However, a nation at war must appreciate and understand the priorities for national defense. Having denied our adversaries the opportunity to successfully wage symmetric warfare, their only other option remains asymmetric. America's military forces are engaged globally, not just against terror, but in a larger context, against the forces that threaten freedom, be they terrorism, drugs, or insurgencies. This asymmetric warfare, by another name, is otherwise referred to as IW. IW is not some ad hoc pick-up game, but rather a conscious strategic choice made by America's enemies to fight by means that expose US vulnerabilities.

Irregular Warfare: Winning the Long War

Irregular warfare requires a particular mind-set and specific talents not entirely applicable or common to more traditional styles of warfare. The relative lack of predictability and its indifferent boundaries regarding what each fight constitutes in terms of objectives and resources are troublesome characteristics well beyond the numbers involved. By definition, IW offers a weaker opponent a viable military option against a stronger one, thereby attempting to thwart America's conventional military strengths. As a result, and by design, IW is warfare where the ostensibly stronger opponent must reframe how it brings its traditional strengths to bear against an apparently weaker enemy.

The USAF has operated with some success in the IW environment before, but has lost significant capacities following draw downs or conversions after each conflict. This should not come as a surprise, given the fact that budgets for unused tools are a luxury not easily afforded in any era. But the extended lead times required to essentially relearn IW each time have significantly affected the USAF's ability to contribute early and effectively in each IW fight.

In developing an effective IW strategy, the USAF must first change its strategic objective from successfully waging IW *to enabling a partner to fight IW*. In the absence of other alternatives, the USAF may serve as a fighting force, but even at that point, the service should adopt the strategic mind-set that it is conducting a holding action while the supported partner builds/enhances its own capabilities. Winning strategies in IW are conducted by, with, and through the supported partner. Furthermore, barring annihilation options, no substantial history exists to support the idea that any outside power can win an inside war. The *Quadrennial Defense Review (QDR) Report* of 2006 provides an important framework for this discussion:

> Long-duration, complex operations involving the US military, other government agencies and international partners will be waged simultaneously in multiple countries around the world, relying on a combination of direct (visible) and indirect (clandestine) approaches... Maintaining a long-term, low-visibility presence in many areas of the world where US forces do not traditionally operate will be required. Building and leveraging partner capacity will also be an absolutely essential part of this approach, and the employment of surrogates will be a necessary method for achieving many goals.

Despite the recent publication of Air Force Doctrine Document (AFDD) 2-3, *Irregular Warfare*, and AFDD 2-3.1, *Foreign Internal Defense*, the USAF has not yet established the institutional and systemic changes necessary to implement its doctrine for operational art in the Long War. Pending such change, the USAF continues to view IW as a subset of peer or near-peer conflicts—a fundamentally flawed position. Solutions in IW must be based on flexibility to address the unique local circumstances of each conflict, where solutions are largely generated from political rather than military initiatives.

If the USAF wishes to become effective in IW, it should acknowledge the necessity of fulfilling two sometimes competing missions. First, and in the absence of alternatives, **the service should have full capability to engage an asymmetric enemy directly—to fight IW engagements as a key component of the national effort**. This will inevitably occur in a joint, interagency, and coalition-based context. Second, **the USAF's most fundamental job is to help establish a credible host-nation air force.** The USAF should have the capacity to create within a partner nation the requisite skills and disciplines in air, space, and cyberspace that enable that partner to realize its national goals without the large footprints or heavy hand of US military presence. Today, that transfer/training capability exists on a permanent basis in only one relatively small squadron in Air Force Special Operations Command: the 6th Special Operations Squadron (SOS). The 6th SOS certainly has the talent, but lacks the required mass for engagement and persistence across the breadth of areas affected by the Long War. The USAF's general-purpose forces have the necessary mass, and with the appropriate force-development program, can have the requisite talent as well. **The USAF should step up to the challenge and shape its force for a winning strategy in the Long War that will enable Airmen to *assess*, organize, train, equip, *assist*, and *advise* foreign air forces for success against irregular adversaries.**

In ensuring air superiority, the USAF's fifth generation fighters are a key ingredient. They are required to make entry-level opportunities available for both the current and future fight. The F-22, combined with the unrivaled capabilities of Airborne Warning and Control System (AWACS), F-16s, F-15s, Joint Surveillance Target Attack Radar System (JSTARS), Navy F-18s, and soon F-35s, will provide air superiority that creates a requisite freedom of action for US and coalition forces. However, with an increasing acquisition cost highlighted by the recapitalization challenges, the numbers of aircraft needed to support all theaters is problematic. Future budget strains will place at risk aircraft, space, and the cyberspace systems acquisition—and personnel—necessary to support future conventional campaigns. What is required, then, is a **"right-tech" solution that meets both immediate USAF IW requirements and provides an enhanced capability for partnering with host air services**. This is not a "low-tech" alternative, but rather providing the right technological solution appropriate for the situation. For example, in rebuilding the Iraqi and Afghani air forces, it will be years before they are prepared to support even the most basic F-16, and even then, it might not be the best platform for their purposes.

The USAF must consider procuring aircraft specifically designed for IW operations to augment and balance the current force—the same aircraft that best augment and balance a partner's force. New, highly capable right-tech aircraft, operating beneath the top cover of air superiority, can be acquired at lower cost and in greater numbers than more expensive state-of-the-art aircraft. **These new right-tech aircraft also provide the opportunity for platforms that can be assimilated by host-nation**

forces, creating capacity where none may have existed before. This strategy will enable partnering for decades through bilateral agreements as the United States provides logistical and advisory support at levels as desired and appropriate for both nations.

Building partnership capacity is not a particularly novel idea. Many coalitions and alliances have formed around similar concepts; in the North Atlantic Treaty Organization (NATO), for example, "interoperability" initiatives provided essentially the same effect. Within the proper context, this approach not only extends military tactical and operational proficiency, but it also lends itself to the strategic goals of building and developing a partner-nation's central-government legitimacy and credibility.

Such an enabling strategy produces the immediate benefit of significantly reducing the profile of Americans in contested areas. As seen recently in Iraq, and repeatedly in history, large-power footprints can become a significant rallying factor for traditionally disparate, antigovernment groups. History readily demonstrates how even groups that would never work together in normal circumstances have joined in efforts to eject outsiders. Rather than helping a partner's government, a large US footprint can become a force multiplier for insurgent recruiting and propaganda. An enabling strategy with the proper emphasis on *by, with, and through*, the host government diminishes the risk of the US footprint working against it.

As stated previously, the Long War presents a different set of needs and competencies than peer adversaries or conventional warfare. Thus, if the USAF desires capability in this sense, its force-development system should produce people qualified for IW, as well as more traditional styles of conflict. Force development is a balance of three core efforts—education, training, and experience—designed to ensure that the USAF has qualified people in place at the right time to fulfill the assigned missions. The USAF cannot confine IW to a single specialty or set of specialties. **Force development, thus, should prepare every specialty, though perhaps not every Airman, for IW.** The battlespace is global in every sense, and every medium is affected.

The 2006 *QDR* "quad-chart" identified four distinct patterns of conflict for which the United States must prepare—traditional, irregular, catastrophic, and disruptive. A concern is that forces optimized for one type of conflict will not be prepared for other options. Concurrently, realistic budget constraints preclude separate forces for every possible corner or blend possible in the quad-chart. The challenge, then, lies in balancing military forces capable of responding to the nation's needs. There are several venues for innovation within this environment. For example, F-22 pilots will conduct much of their training in simulators due to the complex and extensive nature of the capabilities within each platform. **With a companion trainer aircraft available for wider application, pilots could be operationally ready to fulfill multiple missions.** The dual qualification would provide airmanship and growth for the crews at much lower flight-hour costs. **Additionally, if the companion aircraft were IW suitable—intel-**

ligence, surveillance, and reconnaissance (ISR); light attack; battlefield mobility; etc.—a ready reserve for IW, including conventional campaign phase IV, would be available. A likely home for such a program would be the Air National Guard, where disaster relief is closely related to many IW functions. What governor would not be interested in an on-call ISR function for initial assessment and reactive updates in a time of need for crisis management from natural causes or a terrorist attack?

Air Mobility

Air mobility, both strategic and tactical, makes global reach possible, enhances global vigilance, and provides the ability to project global power. No nation in the world can respond as quickly and with as much capability as the United States—air mobility makes this possible. The challenge is that USAF mobility must be able to ensure global access across the spectrum of conflict, while remaining both efficient and effective.

Strategic airlift assets, while exceptionally capable, are wearing out. The C-17—with 18 pallet positions—is the workhorse of the strategic mobility fleet and is performing admirably under a higher-than-planned operations tempo environment. The C-5, with greater overall lift capacity—36 pallet positions—is experiencing a mission–capable rate well below 70 percent. Planned reliability measures *via* the Reliability Enhancement and Reengineering Program will improve the performance and reduce operating costs, but these modifications will address less than half of the C-5 fleet. Even with the programmed modifications, the strategic lift capacity of C-5s and C-17s cannot meet the requirement of 33.95 million ton miles a day as stated by the Joint Requirements Oversight Council. **The USAF must seriously consider extending the line for additional strategic airlift platforms beyond those currently planned.** Mobility assets cannot afford gaps in airlift capacity caused by acquisition delays or maintenance-intensive airframes.

Strategic airlift in the future must leverage new technology to enhance their ability to deliver effects. Current strategic airlift strategy is tied to fixed-wing airlifters that require tankers to enable rapid deployment direct to their destination. **The USAF should explore the potential to employ modern airships to bridge the gap between rapid fixed-wing airlift and fast-sealift platforms.** These lighter-than-air vehicles, with their ability to carry 500 tons of supplies or equipment, provide an economically efficient and relatively timely transport. These airships would alleviate some long-haul missions from the C-5 and C-17 fleet, particularly missions within non-hostile environments—like the continental United States (CONUS) or overseas—into no-threat environments. This construct would allow cargo and personnel to arrive at their destination within days, rather than weeks, at a substantially reduced cost.

In addition to providing improved strategic airlift capabilities, the USAF must also address future intratheater airlift deficiencies. Unlike other USAF assets, the general concept of operations for airlift, as it relates to conventional operations, IW, and humanitarian relief efforts, is not significantly different on a tactical level. However, the differences in these operations make it difficult to craft an airlift strategy that is optimized for any particular operation. The US Army's transformation may create the greatest impetus for changing the requirements of a future mobility force. As part of a jointly interdependent force, the operational concepts of airlift must evolve in concert with those of the Army. At the same time, USAF airlift must maintain the ability to support operations during major combat operations.

The current intratheater airlift platform has proven its worth on numerous occasions, yet many of the current assets are nearing the end of their useful life. Asymmetric and nonlinear warfare create a need to support smaller, more dispersed combat units. Current operations in Iraq and Afghanistan have highlighted certain inefficiencies inherent in operating an airlift platform not specifically designed for the new generation of warfare. The C-130s have been a true success story for the USAF, aiding the warfighter while alleviating some ground transport needs. However, the USAF has often flown C-130s at less than full capacity. A 500-nautical mile intratheater requirement of two to three pallets falls well short of the C-130's 2,000-nautical-mile range with a normal load of six to eight pallets with approximately 35,000 pounds of capacity. In addition, smaller dispersed operations made it more difficult to find a suitable airfield near the war fighter. The C-130 can operate out of approximately 5 percent of all airfields in South America and 15 percent of airfields in Africa—only 207 airfields on the entire African continent are suitable for C-130 operations. **The intratheater airlift assets of the future must be capable of operating from short, austere, unimproved airstrips while maximizing small unit lift requirements to facilitate their usefulness throughout a distributed combat environment.** At the same time, the USAF should strive to field an airlift platform that can be sold *via* foreign military sales to enhance our allies and coalition partners, while enhancing long term partnering requirements of emerging nations.

Certainly key to mobility projection is access to airfields and infrastructure. If the current trend of CONUS-basing continues through 2018, options for basing, ramp space, and runways also become more tenuous. Overseas basing reductions must be balanced with aircraft capabilities as well as strategic considerations as the nation moves forward. As the forward-deployed locations become more austere and less numerous, considerations must be given to how we will insert forces and sustain them during combat.

A successful air-mobility operation depends on a network of facilities, diplomatic clearances, and usable destinations, which include airfields and drop zones. Access to theater airspace and airfields throughout the world presents a major limiting factor to air-mobility operations. Throughout the emerging world, aircraft are often forced to use austere airfields. These airfields are limited by runway condition and size, taxiway systems, and ramp space. To overcome these restraints, the USAF should consider the use of short takeoff or landing and vertical takeoff or landing aircraft with shorter runway and parking needs. Estimates show these aircraft could operate from 80 percent of airfields in Central and South America, in addition to use for disaster operations at home.

Strategic Communication: Spreading the Word

A successful strategic communication strategy is critical to the USAF's success in the global battlespace. However, to date, the nation has proven itself to be less astute about strategic communication. For any strategy to succeed, the nation must recognize that communication takes place in a globalized, interconnected environment—an environment in which our adversaries are quite competent and effective. A significant reason for this deficiency is that in a democracy there are often numerous competing agendas at play, generating multiple competing narratives. Often, these inherent complexities force the USAF to continually be in a reactive mode.

To develop a broad service strategy, the USAF must collectively, persistently, and convincingly communicate its unique global effects–generating capability. This communication strategy within a service strategy is almost wholly dependent upon remedying an internal Air Force cultural reticence to communicate for effect at the most senior levels. **Consequently, the USAF must develop information and influence operations as a complete weapon system.** For example, public affairs (PA) experts—in collaboration with information operations (IO) and public diplomacy experts—must integrate at every level of operational planning and execution. The key to successful strategic communication lies in understanding that desired effects should drive the targeting option rather than the other way around. Consequently, influence operations, as a weapon alone, may constitute the best means of affecting a target.

As a long-term goal for force development, the USAF must ensure that IO experts are better developed and sustained. For example, PA officers, who have long been stove-piped as specialists in their career field, should broaden their understanding of related areas. One option might entail combining PA and intelligence career fields with alternate assignments. Another option, in light of proposed cuts in the number of PA officers, might involve re-creating additional-duty options for operators as PA officers. Built around a structure of limited–length tours with a front-loaded training course, the option would give individual wings a PA officer chosen by that commander and, over time, would significantly increase the number of line officers with a better understanding and appreciation for the information environment and the effects that can be generated in that environment.

Culture and Language in the Expeditionary Air Force

Global Vigilance, Reach, and Power require culturally, regionally, and linguistically competent Airmen who will exert positive influence in support of Air Force expeditionary operations. On any given day, over 33,000 Airmen are engaged in expeditionary operations around the globe, conducting conventional and IW; stability, security, transition, and reconstruction; humanitarian relief; Foreign Internal Defense; and coalition building. Culture, region, and language efforts enable their success.

Language offers the gateway to understanding culture and serves as a critical first step in appreciating a center of gravity—the local population. DOD initiated progress via the 2005 *DoD Language Transformation Roadmap*. In 2006 and 2007, the USAF followed suit, establishing the Air Force Culture and Language Center at Maxwell AFB, and adding culture and language familiarization to the Air Command and Staff College (ACSC) and Air War College (AWC) curricula. Additionally, the USAF Academy added two new language offerings and increased its language requirements for non-technical majors. Air Force Reserve Officer Training Corps created more opportunities for language immersion for its cadets. The Air Staff moved language under Headquarters Air Force (HAF)/A1 and the DOD significantly raised foreign language proficiency pay.

The Air Force Culture and Language Center was designated to coordinate education, training, and research and to ensure that USAF-wide efforts were guided by subject-matter experts who could leverage the efforts of other USAF and DOD institutions as well as the larger academic community. The strategic goals of the Air Force Culture and Language Center are to ensure: (1) the larger strategic **development of cross-culturally competent Airmen**; Airmen—with appropriate levels of cross-cultural knowledge, skills, and attitudes—who are able to meet Air Force mission needs and able to surge for emergent requirements. (2) **Alignment of Total Force cross-cultural competence capabilities with requirements** and Air Force institutional competencies in the Total Force aligned with operational requirements to support Joint and Air Force missions. This will help Airmen understand and influence operations, activities, or actors—joint, interagency, allied, coalition, noncombatant, and adversary. (3) **Sustainment of Total Force cross-cultural competence capabilities**; agile, responsive, and cost-effective plans and policies that sustain Air Force cross-cultural competence capabilities.

However, while addressing the cultural initiatives within the USAF, none of these actions resolved the issue of language proficiency for Airmen. Except for the cryptanalytic career field, no USAF language requirements exist. Only officers in the International Affairs specialist program or attachés—individuals normally at the rank of major and above—truly have the opportunity to use their language skills. Even once developed, the USAF still does not systematically ensure anything beyond basic language or culture skills is maintained.

Initial guidance, such as the 2005 *DOD Language Transformation Roadmap*, provides a vector for USAF efforts. Language will remain a critical component of learning for specialists. However, determining a strategy for the next 10 to 15 years, and beyond, is a challenge. Who would have predicted in the spring of 1989 that in a few short years Mandarin Chinese would be a language in demand and Russian would fall out of vogue—though it may be making a comeback. **Despite not being able to predict every future contingency operation, the USAF must determine its requirements for languages and plan accordingly. It must respond to the *DoD Language Transformation Roadmap* with a comprehensive, sustainable language "flight plan,"** encompassing all disciplines and emphasizing a long-term commitment to language education and training. **Accordingly, in a break from past practice, force-development planning for languages and culture should also include a sustainment package to ensure future resource proficiency and availability.**

Identifying Airmen with language skills is only the first step. The USAF must also *create* Airmen with language skills. It takes effort to acquire some passable fluency in another language and a great deal of perseverance to maintain it. One approach to resolve the language issue is for all incoming USAF officers to possess a language proficiency as a prerequisite to being commissioned. This will promote formalized language knowledge throughout the officer corps, albeit at its lowest tier and with varying degrees of competence. However, currently few students in the science and engineering fields complete degree requirements in four years. An additional language requirement would further lengthen the time already spent by those students in school. Given the USAF's need for technically proficient Airmen, this becomes a self-limiting condition.

The intelligence community has recently considered requiring all intelligence officers to have language skills, thus providing an excellent cohort to implement a new approach; one where each intelligence officer "becomes a language." **To produce enough linguists in a variety of languages, each newly minted intelligence officer would be assigned a language from the DOD directive's Strategic Language List. For their entire career—including Air National Guard or Air Force Reserve duty—it would be their professional responsibility to maintain proficiency in a language.** Further, given that the overwhelming majority of linguist billets reside in the intelligence community, **the USAF should consider moving the service program manager for language from HAF/A-1 to HAF/A-2**. In this way, Lieutenant Jones would become "Mandarin Chinese" and Lieutenant Smith would be "Arabic." While maintaining such proficiency would be an individual professional responsibility, USAF programmers and managers must provide time and resources to support such individual efforts as USAF "weapon systems." If and when needed, this ready reserve of language-capable officers would be proficient and available for use by the USAF and greater DOD community.

Distributed Planning: The Key to Centralized Control

To effectively execute Global Vigilance, Reach, and Power, USAF doctrine states, "Centralized planning is the first step to achieving effective centralized control." This statement, while true in traditional combat operations, creates a construct across the full spectrum of warfare that is less effective. As the USAF finds itself involved in nonlinear, asymmetric warfare consisting of simultaneous and continuous offensive, defensive, and stability operations conducted in a complex, highly integrated, networked, and distributed environment, centralized planning proves less effective. Additionally, these operations will most likely take place under the control of a joint task force (JTF) commander. This environment requires ground units at the tactical and operational levels of war to operate predominately in distributed areas of operation, thus forcing their planning to the tactical level to take advantage of individual and small unit initiative (Field Manual 3-0, *Operations*). What this means is that while centralized control remains paramount to retaining theaterwide effectiveness, **to effectively integrate USAF capabilities with joint operations at the JTF level and to impact air and ground planning at the earliest stages, the USAF must adapt to a distributed planning model to maintain effective centralized control.** For without clear, distributed joint planning there can be no meaningful prioritization—all of which enable effective centralized control. **Distributed planning dictates that the USAF must place experts with the appropriate planning tools at the locations where operational plans are born and refined.**

In most geographic commands, the USAF has one Air Operations Center (AOC) to support the entire area of responsibility (AOR). Under the current planning construct, the USAF has consolidated the majority of its expertise—air, space, and cyberspace—in the AOC. When a JTF is created in-theater, the USAF is not positioned to send a joint forces air component commander (JFACC) supported by an AOC to that JTF. In an attempt to create a presence, the USAF has chosen to assign an air component coordination element (ACCE) to JTFs. While an ACCE provides air expertise at the JTF level, it is not involved in the formal planning process, either with the JTF or the AOC. The end result is the JTF staffs do not have adequate airpower planning expertise or the organizational "hooks" into the formal joint air operations planning process within the AOC. This situation is magnified if multiple JTFs are created within a combatant command. If the USAF ever hopes to impact planning at the JTF level, it must develop an organizational construct that adequately provides a JTF staff with distributed planning expertise.

At the JTF level, the current ACCE construct can provide the expertise needed to solve the JTF-level planning gap, but certain changes will be required. **The USAF must identify individuals with the appropriate expertise to serve in the ACCE and track them in the personnel system.** These members should possess the expertise needed to apply the full range of USAF capabilities to support a potential JTF. **Second, the USAF must provide qualification and currency training so they are prepared to support a JTF when one is established.** JTF ACCE modules could be pre-built using unit type codes, expediting the deployment of ACCE qualified teams.

In addition, operations require planning expertise at organizational levels below the JTF. The US Army has pushed planning for ground operations to lower levels—brigade level and below—in order to enhance initiative and respond to the specific nature of local operations. However, USAF planners do not reside at commensurate levels. The USAF's standard planning construct thus places Airmen at the end of the planning process, not at the beginning where they can be most effective. **For effective planning, the USAF must have planning expertise at the brigade-level equivelant and below.** Individuals involved in air-to-ground strike planning and execution do have a presence below the AOC level; however, their expertise resides in close air support (CAS) and terminal control of aircraft, not the planning expertise needed to support the dynamic nature of distributed operations across the full spectrum of operations.

To address the air and ground integration challenges below the JTF level, the USAF should establish a new unit that is organized, trained, and equipped to conduct distributed planning. Further, experienced planners should be assigned on a permanent basis to this unit. This organization can be supplemented with additional personnel through the Air Expeditionary Forces (AEF) process, but the core cadre should be permanently assigned trained professionals. The USAF can leverage the recently established Air Ground Operations Wing at Moody AFB to create this organization. Through developing subordinate group and squadron size structures, the USAF would create organizations that should align to the US Army division, battalion, and company levels. These permanent structures would replace the *ad hoc* nature of today's organization used to support the noncontiguous fights in Iraq and Afghanistan.

Even in a resource-constrained environment, there are several ways to provide the necessary manning to support the new distributed planning process. With a greater degree of the planning done at lower levels, fewer personnel will be needed within the AOC, thus freeing up support for distributed planning. Out of this available pool of manning, some slots should be designated for planning support at the ACCE/JTF level. These individuals, available to work in the AOC on a day-to-day basis, must be pre-identified for ACCE duties and sent to a JTF when needed. The remaining slots would support the new planning units below the JTF. Further, as there will not be a one-for-one replacement as the USAF recapitalizes its air assets, personnel associated with the legacy systems could be retained to support distributed planning. These individuals possess the core expertise to ensure distributed planning success.

Finally, to ensure the proper integration and synchronization of air, space, and cyberspace power, the USAF may have to make resourcing distributed planning a priority, despite austere budgets. Accordingly, if manning shortfalls from the AOC and recapitalization are inadequate, the Air Force may need to invest in the additional personnel needed to grow the required planning expertise. In 2006 the Air Force faced a similar challenge. Due to US Army reorganization and the distributed nature of IW occurring in Iraq and Afghanistan, there was an increased demand for joint terminal attack controllers (JTAC). Despite ongoing personnel cuts, the USAF decided that the CAS mission was critical and met the challenge by increasing the JTAC career field by approximately 900 Airmen.

Unmanned Aerial Systems: The Air Force in Transition

Long before transformation became a catchword for forward thinking, the USAF proved itself as a transformational service. During the early years of the Cold War, the USAF transformed itself from a conventional bomber force to a global strike force responsible for nuclear deterrence and national defense. Also, during that same era, the USAF embraced a second domain: space. That evolution continues today as the USAF begins to tackle cyberspace as a third operational domain. In addition, the USAF is facing a new challenge—the evolution of unmanned aircraft. The USAF will continue to have a need for manned aircraft, as manned systems provide flexibility not available through technology alone. However, the question before the USAF is how will it respond to fully embracing a myriad of new, unmanned weapons system.

Today, unmanned aerial systems (UAS) are reaching the operational capabilities of both air and space assets. Commanders use UASs in Iraq to surveil well-traveled convoy routes, keeping an eye out for improvised explosive devices—a task once requiring that soldiers sit in Humvees for hours on end. Platoons use model-airplane sized UASs, such as the Army's Raven, to track enemy movements from house to house. Systems like Global Hawk and Predator are playing increasingly important roles in all IW operations. They provide the capability to observe an adversary sight unseen, deliver live-action video to a command post in-theater, and can precisely destroy a chosen target; all while being flown from halfway across the globe. Integrating networked capabilities through air, space, and cyberspace using Global Positioning System (GPS) and satellite communications, UASs embody all aspects of Global Vigilance, Reach, and Power—but, of course, they do so only in a permissive environment guaranteed through air superiority.

The USAF is at a transformational crossroad, yet again. The challenge is more than bringing a new weapon system into the fold; it is about developing a new cultural construct within the USAF. Unlike space and cyberspace, UASs strike at the heart of long-held cultural beliefs among the service's core community, which is that officer-pilots solely fly, fight, and win in the air. Further challenging this core cultural belief is the fact that the next evolution of UASs is well within the timeframe of this study; that is, UASs that may be space-based, supersonic, hypersonic, and/or controlled by air-breathing platforms flown by pilots. Ultimately, the next-generation bomber—the one beyond 2018—may well be an unmanned system.

How, then, should the USAF transform itself, once again, to take full advantage of UAS operations? First, **the USAF should develop a separate operational career field, with a separate qualifying weapons school**, not unlike Undergraduate Pilot Training (UPT) or Undergraduate Space and Missile Training (USMT) where young officers learn the basics of the profession. This new operational career field would have a cross-flow from other operational career fields within the USAF, where dual qualifications would be encouraged, with primary and secondary Air Force Specialty Codes (AFSC) designating capabilities that would enable the USAF to track expertise.

During this study's timeframe, it is highly likely that UAS force development will echo what occurred in space operations. During the early days of space exploration, satellite operators were engineers. Following the USAF's effort to "operationalize" space, officers with degrees in history and forestry became satellite operators. Today, satellite operations are accomplished by a team, with enlisted personnel on the consoles conducting the uplinks and officers "in the loop" when critical operations are required. In the next decade, the USAF will find it desirable to fully incorporate the enlisted force into "flying, fighting, and winning the nation's wars" as UAS operators. Following the space operations example and using a team construct, enlisted, contractors, or DOD civilians would likely fly the UAS, but when weapons release is required, a UAS rated officer would participate in the decision making/execution loop.

Currently, Air Combat Command (ACC) is the home for the UAS—and this is appropriate for the current fight. ACC understands the strengths of the current generation of UASs—air-to-ground and air-to-air operations—and is best positioned to take advantage of the benefits UASs bring to the combatant commander. With time, UASs will evolve as transport aircraft and once that occurs, Air Mobility Command (AMC) is the right command to control those assets, just as it does with manned airlift.

As the USAF adapts to new UAS roles, it will also find a need to robustly integrate manned and space systems into a coherent networked structure. This network will include all services operating both ground and sea-based systems, effectively exploiting the human-machine interface as a critical part of capitalizing UAS capabilities. This construct will enhance operations across all domains as integration and interdependence become key factors in the success of the entire system.

Another area yet to be fully explored involving UASs is the arena of homeland defense. Following Hurricane Katrina in 2005, the United States used satellites and low-level aircraft to assess the extent of damage to the Gulf Coast. While the military provided significant assistance during the days following the disaster, DOD could not provide UAS coverage of the affected areas. Since then, the Federal Aviation Administration (FAA) has approved UAS over-flights within the United States. However, only FAA licensed pilots are allowed to operate UASs within US airspace. **The USAF should take the lead in determining common language and jurisdictional procedures between the Justice Department, the FAA, and DOD**

to enable the widest possible use of UASs at home. Once resolved, the USAF will be able to provide incident awareness and enable assessment while continuing to protect individual civil liberties.

Finally, while UASs can be controlled from thousands of miles away, they must be commanded locally and be accountable to the theater commanders, following the concept of centralized control and decentralized execution. Accordingly, **UAS operational missions—such as reconnaissance, strike, and electronic warfare—should be apportioned in theater, accountable to the JFACC, and then defined doctrinally in the USAF and Joint communities afterward as a statement of best practice.** Consequently, the USAF need not control all UASs, as it has no desire to control all other air assets. Service-appropriate UASs, such as the Army's Raven or naval carrier task force assets, fit within the organize, train, and equip realm of their respective service. However, DOD assets with theaterwide application—like Predator, Global Hawk, and the US Army's Warrior—should be USAF-controlled and apportioned systems *via* theater-assigned JFACCs.

Intelligence Reform: The Secondary Effects of Merging Intelligence with Surveillance and Reconnaissance

ISR is at the core of the modern USAF. From the balloon corps of the US Civil War to the earliest flights of biplanes across the trenches in World War I, air assets have been used to gather intelligence, conduct surveillance, and "recon" the battlefield. However, no period since Pearl Harbor has resounded with calls for intelligence reform like the present. A recurring theme in the USAF has been to "tighten the kill chain;" (*i.e.*, reduce the time between the sensor and the shooter). This puts the burden on the intelligence community to produce "actionable" intelligence as quickly as possible.

Technical innovations are driving questions such as, "Is intelligence now operations, and if not, is there a real difference?" Sensors and shooters are often one in the same—think Predators armed with Hellfire missiles. In fact, former Secretary of the Air Force Michael W. Wynne, stated at the Air Force Association Air and Space Conference held in Washington, DC, 25–27 September 2006, "Every sensor will be a shooter, and every shooter will be a sensor, linked across all domains and across the joint and the coalition team." Highlighting the USAF intelligence "challenge," Gen T. Michael Moseley stated:

> Through technological advances and Airmen's ingenuity, we can now surveil or strike any target anywhere on the face of the Earth, day or night, in any weather. A more challenging issue today—and for the future—is determining and locating the desired effect we want to achieve. Because ISR capabilities are at the core of determining these desired effects, ISR has never been more important during our 60 years as an independent service. ISR has become the foundation of Global Vigilance, Reach, and Power. (*Air and Space Power Journal*, Summer 2008)

One senior USAF officer noted that "ISR is an operational function with the goal of providing accurate, relevant, and timely intelligence to decision makers." This is consistent with the approach that ISR is operations. However, care must be taken if the USAF attempts this major doctrinal shift. Intelligence is a product and not a subset of the platform—thus while an F-15 may gather intelligence, it is not an intelligence platform. On the other hand, while a Hellfire missile may be fired from a Predator, at the end of the day the Predator remains an ISR asset—thus flexibility of use and not the platform, as such, should remain the focus.

The assertions that ISR is indivisible and that ISR is operations are both true, but the USAF must take care how it defines ISR as a single operational function in doctrine. Intelligence relies upon surveillance and reconnaissance activities, but each demands a specialized focus to ensure the warfighter has the best possible integrated product. Without analysis, surveillance and reconnaissance data remain just that, data, and never cross the threshold to become intelligence. Consequently, **ISR should not return to the A3 community (*i.e.*, USAF/A2/3), but certain common functions below the A2 and A3 should be linked together to foster effective management, oversight, and fusion.**

A key challenge facing the USAF intelligence community is that while it has experienced significant change over the past 15 years, much of which has made it less responsive to the warfighter. As a short primer, the USAF eliminated human intelligence (HUMINT) and targeting as an AFSC in the 1990s. Additionally, signals intelligence and air intelligence officers merged their training at Goodfellow AFB to become one renaissance-type intelligence officer. The end result was a de-emphasis on HUMINT and target analysis, while at the same time training time increased for intelligence officers as the career field attempted to create a "one size fits all" system.

The benign neglect and reliance on intelligence officers, at large, to accomplish the highly specialized tasks of targeting and target analysis have left the USAF with a gap in its analysis capabilities. At a time when collateral damage must be minimized and target selection becomes part of the strategic-communication war, specialized expertise is a must. Analysis and interpretation have become a lost art, driving many intelligence officers to be little more than readers of Defense Intelligence Agency (DIA)–derived data. **Accordingly, the USAF should create a separate targeting career field for Air Force and Joint targeting support. Additionally, to further address this deficiency, the USAF must emphasize analysis within the sensor-to-shooter chain to assure actionable intelligence support to decision makers.** Simply, with increasing technological advances, the sheer weight of information/data to be rationalized has increased exponentially.

When DIA took over the HUMINT mission for the USAF in the mid-1990s, it promised to meet USAF HUMINT requirements. For numerous reasons, DIA has not followed through on its promise. Consequently, following the attack on 9/11, the Air Force Office of Special Investigation (AFOSI), using existing resources, took on the USAF HUMINT mission in the Central Command (CENTCOM) AOR. **To correct the loss of capability within the USAF, it must reestablish HUMINT as an AFSC. However, as it will take years to become viable again, the USAF must, in the interim, resource the OSI to continue their CENTCOM mission for the near term.**

Another area of increased emphasis must be intelligence support to the flying, space, missile, and cyber units and their use as a training ground for all intelligence officers. **The USAF should re-energize unit intelligence assignments as a building block for understanding support to the war fighter.** The first assignment for *all* new graduates from the intelligence school should be at the operational unit level. Nothing provides professional growth and personal perspective for a new lieutenant as briefing a mission to a seasoned group of operators. Both gain appreciation for the other's task and intelligence officers understand what operations are about and what intelligence support is required for joint mission success.

The Total Force in Transition

Since the inception of the Total Force policy in the 1970s, the USAF has become the model for Reserve and Guard integration, far outstripping any of the other services in integrating, synchronizing, and using Total Force capabilities. Successful, enduring USAF operations for the past three decades would not have been possible without Total Force contributions. In the past, the Guard and Reserve operated under the Cold War constructs of "tiered readiness" and "Strategic Reserve," terms that in today's lexicon equate to the "weekend warrior" stereotype and resonate with images of old equipment and unready Airmen. The more current construct of "Operational Reserve" succinctly describes the current state of readiness and participation. However, how long can this Operational Reserve continue at its current utilization rate?

At issue is the sustainability of this high operational tempo by the Guard and Reserve over time. The "citizen-Airman" is the USAF's closest link to the American people. However, constant overseas deployments jeopardize the bonds between the two. An unintended consequence of such a high operations tempo is that it has transformed the "citizen-Airman" into the "Airman-citizen."

The Guard's response to Hurricane Katrina was a widely chronicled success story. However, the Guard's increased overseas role since 9/11 creates concerns that it will be unable to respond effectively at home for lack of available personnel and equipment. Calls for a reduced overseas Guard presence continue as states realize that their "first first-responders" may be needed at home sooner rather than later. Additionally, the old argument that it is less expensive to use the Guard and Reserve is valid only if they are not employed on active duty for any considerable length of time, such as in Iraq and Afghanistan. Yet the current method of employment—frequent AEF rotations and use of individuals as "in lieu of" forces—creates, in essence, another active duty bill to pay.

The American people expect their military to respond to any national disaster, man-made or natural, and no one in crisis cares who comes to their rescue or what branch they represent. However, the military is often constrained from providing immediate disaster relief response except in the most extreme situations. The DOD is not designed to be a first responder, nor does the *National Response Plan*, or the newer *National Response Framework*, envision it in that role.

To return the balance to the "citizen-Airman" construct, the Guard's primary mission must become homeland defense, first and foremost; this includes the Air National Guard (ANG). However, those who counsel that any task for homeland defense can be met as an adjunct of wartime training ignore the subtleties of working with state and local governments, all of which have a tribal flavor. The very nature of "consequence management" is designed to help people and to protect property, not to achieve kinetic effects against an armed foe. **Thus, the ANG must place a priority on homeland defense planning in conjunction with state homeland defense authorities and their Army National Guard counterparts.**

Homeland defense planning for the ANG should first start with dual-capability assets—units that have use during a disaster at home and in combat operations—such as medical personnel and Medical Evacuation capabilities. **Accordingly, ANG medical units will require modification to their existing designed operational capability (DOC) statements designating them to perform disaster-relief missions**. In relation to dual DOCing medical units, the revised DOC would provide for unique first-response medical needs beyond the typical emergency medical specialties. For example, there is a new physician certification in disaster medicine that encompasses all phases of disaster medical delivery, which is a natural complement to existing unit medical capabilities. Additionally, psychiatric specialists will also be required to address posttraumatic stress disorder patients as well as "the worried well"—those without apparent injuries who nevertheless seek medical assistance. As a part of a response package, the Guard will need to identify personnel with gerontology expertise, either as a primary or secondary specialty. The aftermath of Hurricane Katrina saw large numbers of nursing home patients abandoned by their long-term care facilities. A similar situation in any large-scale disaster would strain triage of other disaster victims.

In the absence of any explicit DOD planning guidance, and in view of the uncertain nature and timing of natural and man-made disasters, **the National Guard Bureau and the adjutants general of the several states must create disaster-response packages from their existing resources.** In time, downward-directed guidance will be provided, but an opportunity exists for the Guard to shape that guidance today.

Homeland defense DOCs for rescue and airlift should be added as secondary DOCs for both the ANG and Air Force Reserve Command

(AFRC) units. In addition, the ANG should codify its domestic counter-air role within a DOC statement for selected units. In these ANG units not already designated to receive the F-22, retiring F-16s should be replaced by cargo aircraft and other airframes consistent with the homeland defense mission. This is not to cede the Air Guard's fighter mission, there remains a need for the Guard's combat capabilities to meet current operations and any postulated conventional campaign, but rather to augment its existing capabilities. **As current equipment is programmed for retirement in the out-years, selected ANG units should be re-equipped with dual-capable platforms and be assigned DOCs allowing for use in multiple mission areas.**

The Air Force Reserve (AFR) is a viable, vital part of the Total Force, but it must reshape its structure to better manage its resources. **Most AFRC units should continue to focus on combat operations, as they are the backbone of the command and must receive the lion's share of its funding and missions.** The AFRC's Reserve Associate mission should continue in every venue where it is appropriate and where it results in manpower and equipment savings. However, care must be taken in an attempt to gain efficiencies, specifically when it comes to the Guard and Reserve. Mixing AFRC and ANG assets in an Associate grouping would be cumbersome and possibly contentious, pointedly so in a state response to a natural disaster. Structurally, the AFR must realign itself to make better use of its current assets. AFRC numbered Air Forces (NAF) must assume a warfighting role within their gaining organizations or risk being labeled a cold war construct. **The three AFRC NAFs must assume a near-mirror-image warfighting capability like their active duty counterparts, or be downsized or eliminated.**

In an additional attempt to gain efficiencies, the AFR is beginning to use the Individual Ready Reserve (IRR), traditionally not a part of their AFR culture. A shift to this nonpay, nonstructured resource will be difficult. First, the IRR cannot continue to be seen as a large, amorphous "repo depot." They must be assigned against validated wartime billets. This is the *Strategic Reserve*. Next, in order for the USAF IRR to be an effective force multiplier, it must allow its members to train close to home to reduce its personnel's "out of pocket" expenses. Finally, if the USAF desires to use the IRR as a manpower pool, the IRR database and management tools must be consolidated—currently split between Denver and Robins AFB—at a single location, preferably at HQ AFRC. **Ultimately, the Air Force IRR should be energized as the USAF's strategic reserve and resources must be allocated for its management and utilization.**

Air Base Disaster Planning: Contingency Operations

Unlike the plethora of detailed terrorist-threat scenarios, there are no in-depth assessments of an air base's susceptibility to natural disasters. Moreover, there are no plans to relocate or reposition a base's primary mission if its real property is rendered unusable by a disaster. Simply put, the USAF knows more about al-Qaeda's threat to AMC's West Coast strategic airlift assets than it does about the Cascadia Subduction Zone's earthquake/tsunami threat to McChord AFB. Nature has seen fit to destroy or severely damage more than one of these bases, highlighting the one factor missing from these basing deliberations: susceptibility to natural disasters. Therefore, **the USAF should conduct a natural disaster base vulnerability study of all active duty, Guard, and Reserve air force bases and develop clear plans for reconstituting bases that have a high probability of being evacuated during natural disasters.** The study would assess the vulnerability of these bases and determine the likelihood of major damage to each and should be used to determine future weapon system beddowns. Further, scenarios where similar missions are affected must be developed along with plans for mitigation.

The continuing reduction of USAF bases through the Base Realignment and Closure Commission (BRAC) process resulted in a depletion of redundant facilities, (*i.e.*, similar repair facilities, hangars, and civilian expertise) throughout the USAF. Today, USAF bases literally stand as their own "island" of expertise. BRAC did not take into account the possibility that some installations may fall within a "tornado alley," lie on a flood plain, straddle a seismic fault, or operate adjacent to a potentially active volcano. If a base with a unique capability is rendered inoperable following a natural disaster, there are few, if any, bases capable of continuing that particular mission without impairing their own unique work. In a zero-sum game world, one-of-a-kind base functions put USAF missions at high risk.

Prior to the next round of base closures—and there will be future closures—the USAF must impress upon congressional committees the vulnerability of the existing base structure and the need to retain capability. Bases—to include Guard and Reserve—should be retained for redundancy in an era of single-mission air bases. Once lost, bases are virtually impossible to recover. Thus, cost savings become only part of the equation balanced against lost capabilities and increased risk to national defense and homeland security.

Chapter 4

Recasting the Future (2018–23)

It is hard to conceive of any country challenging the United States directly on the ground—at least for some years to come. We can expect that asymmetric warfare will be the mainstay of the contemporary battlefield. . . . Success will be less a matter of imposing one's will and more a function of shaping behavior—of friends, adversaries, and most importantly, the people in between.

—Secretary of Defense Robert M. Gates

The widely acclaimed USAF white paper on *Global Reach—Global Power*, published in June 1990, framed the USAF transition from a nuclear deterrence and forward defense posture born in the Cold War to a more flexible posture predicated on our ability to move forces and employ force anywhere on the globe, as needed. It was a powerful and enduring message for the USAF. The white paper's principal observations regarding the changing environment have remained remarkably consistent with history as it has played out over these two decades.

In the face of this uncertainty, **the fundamental US national security objective remains the preservation of the United States as a free and independent nation, with its people, values, and institutions secure.** That basic objective, and our other national security objectives, are supported by interrelated political, economic, and defense strategies. As Gen Colin Powell, the chairman of the Joint Chiefs of Staff, observed, **"We must remember how we got to this historic turning point in history—our systemic strength and the strength of our allies has gotten us here.** And a crucial dimension of that strength is our well-trained, proud, and ready military force." To maintain this state of affairs, **US forces will protect US security interests by providing the correct balance of military capabilities**—forces capable of maintaining deterrence and protecting our national interests. As we search for that correct balance we can neither adopt the unrealistic assumption that nothing has changed, nor the historically naive presumption that everything has changed.

Capitalizing on this successful strategy document and discussion, USAF leaders followed a similar pattern with subsequent documents on global presence (1995); global engagement (1997); and global vigilance, reach, and power (2000). However, by 2000, the larger message encapsulated in the connected phrase Global Reach—Global Power had been substantially lost. The reorganization of the USAF, resulting in the creation of Air Combat Command (ACC) and Air Mobility Command (AMC), and the increased deployed-operations tempo of the 1990s, provided a near-term focus on "mobility" and "power" and generally related those functions to specific organizations. Additionally, high-tempo operations accelerated the recapitalization crunch for major systems at a time when the DOD budget was in significant decline. Global Vigilance became the advocacy bumper sticker for Air Force Space Command (AFSPC), Global Reach for AMC, and Global Power for ACC.

"Stove piping" the USAF strategic message into specific platformcentric or programmatic statements tended to isolate other USAF operational and support elements and deflated the connections between competencies and acquisition. USAF "strategy" was indeed tied tightly to the program objective memorandum (POM), but not in a holistic, strategy-to-task paradigm.

Over time, the three separate AF strategy elements—vigilance, reach, and power—evolved largely into advocacy statements for specific functions within the framework of USAF contributions to national security. The 2008 *Air Force Posture Statement* did not substantially alter the earlier definitions. As such, an overarching, structured USAF strategy is missing. Global Vigilance, Global Reach, and Global Power, as defined and used today, fall short of the full range of strategic options that the USAF can bring to the fight. USAF mission expansion into cyberspace only exaggerates this disconnect.

Definitions from CSAF White Paper (December 2007)

1. ***Global Vigilance***: Persistent, worldwide capability to keep an unblinking eye on any entity—to provide warning on capabilities and intentions, as well as identify needs and opportunities.
2. ***Global Reach***: Ability to move, supply, or position assets—with unrivaled velocity and precision—anywhere on the planet.
3. ***Global Power***: Ability to hold at risk or strike any target, anywhere in the world, and project swift, decisive, precise effects.

The fundamental starting point for a USAF strategic concept should be a framework in which the entire service sees its value and its contribution. It is interdependence *within* the USAF that is the critical missing element in current AF strategy, and this stove-piping practice diminishes the vital connecting elements. As a service, the USAF has focused, and perhaps over-focused, on the people, systems, and sustainment that can be "purchased" through POM actions. However, as an organization that has adapted an effects-based mentality, the USAF needs to rebalance this focus. By "recasting" the definitions with an effects-based orientation, all USAF functions—agile combat operations, information operations, building partnership capacity, etc.—can be logically integrated within a more cohesive strategic framework.

Proposed "Recasting" of Definitions

1. ***Global Vigilance***: **Situational awareness** required for understanding both the necessity for action and the character of the effects nec-

essary to achieve a revised condition or end state. It is the underpinning element, the foundational starting point for Global Reach and Global Power. Global Vigilance includes both the collection capabilities and the cognitive processes necessary to create that situational awareness. It identifies linkages and assesses the potential for success. Further, it validates actual success or progress achieved through the application of Global Reach and Power. Global Vigilance depends on Global Reach to gather data and disseminate intelligence.

2. ***Global Reach***: **Operational access**—connectivity to the objective through air, space, and cyberspace. Global Reach takes varied forms and depends upon circumstances. For instance, it could be a space-based line of communication, an air route for a C-17, or a portal-to-portal Internet connection. Global Reach is interdependent with the other pillars. For example, Airmen must first secure control of—or the necessary degree of superiority over—its principal domains of air, space, and cyberspace in order to exercise Global Reach.

3. ***Global Power***: **Ability to create and sustain effects** through air, space, and cyberspace. These effects encompass a full range of kinetic and non-kinetic, lethal and nonlethal, and constructive and destructive options prosecuted through air, space, and cyberspace either individually, or more likely, *via* a synergistic, mutually supporting campaign. Cross-domain integration of operations in air, space, and cyberspace enables greater speed, precision, and reliability than those restricted to a single domain. This synergy produces the capability for more discrete effects with proportionally less collateral damage and fewer counterproductive effects.

One of the great enablers for achieving strategic ends is strategic partnering. Partnerships as collaborative relationships built on common goals enhance the United States' ability to achieve objectives while simultaneously aiding our partners to achieve theirs. Partnering includes a broad range of activities with both state and non-state entities and may effectively act through or be formally recognized as a coalition or alliance. Partnering may include traditional military-to-military activities, like foreign internal defense or international military training and education programs. It could also be cooperative engagement with nongovernmental organizations during a humanitarian assistance mission, or with an academic institution working to increase understanding in a social field, or new findings from scientific or technical experimentation. **This burden-sharing, across a nearly unlimited array of partners, increases the opportunity for achieving US objectives while concurrently decreasing risk.**

Traditional values of the Airman's art, such as speed, range, and precision, are frames of reference that transfer easily between air, space, and cyberspace. These domains, because of their near-boundless natures, also foster the Airman's global perspective—a natural extension of the airpower legacy—from providing vantage just beyond the nearest hill to transiting continents, seas, space, and cyberspace. **The USAF is uniquely suited to conduct synergistic operations more than any other service across the global commons of air, space, and cyberspace, thereby generating strategic effects that protect the nation and serve America's interests by means of *Global Vigilance, Global Reach*, and *Global Power.***

Chapter 5

The Future Fight

Maintaining USAF Technological Superiority

The twentieth century was an age of technological innovation; historians point to it as the American Century, as the nineteenth had been for the British. From mass production to placing a man on the moon, to "inventing" the Internet, the United States led the way. Innovation made the United States the world's breadbasket, feeding Europe and Asia following the Second World War. Students from across the globe flocked to US colleges and universities to pursue degrees in mathematics, science, and engineering.

Some historians have postulated that the twentieth century was also an era of air and space, and that American dominance derived from its ability to exploit both. To date, the twenty-first century appears to be one of information and communication. Nations such as India, China, and Indonesia seem to be reaping the benefits of understanding a new domain, that of cyberspace. Today, United States citizens are in the minority in the degree fields so necessary to exploit the new domain. The task before the nation, and more specifically, the USAF, becomes how to maintain the unrivaled technological superiority it enjoys today. To do so, it must first reinvigorate an acquisition process that was designed during the postindustrial age, an era that allowed for a 15-year cycle for systems acquisition. Today's era of rapid change and mass information does not allow the United States to remain a world leader in advanced technology unless systems can be acquired and fielded much more rapidly.

Acquisition Reform: Fixing the Process

Most would acknowledge that the DOD acquisition process requires reform. This is not a case where the entire system need be reinvented, but rather redirected. In total, the acquisition process has produced numerous successes over the past 50 years. It has enabled the USAF to sustain a vast technological lead over its nearest competitors in the medium of air, space, and cyberspace. It has fielded such marvels as the world's most capable fighter in the F-22A and space launch vehicles, such as the Atlas V and Delta IV, that will assure US access to space for decades to come. These programs, however, came at a cost. In a time of budget constraints, acquisition reform must occur to ensure the process becomes more efficient, effective, and responsive to guarantee the USAF remains the world's preeminent air force. Thus, the USAF **must advocate to DOD and policy makers means to shorten the acquisition cycle,** reduce cost, and take full advantage of technological innovation.

In March 2008, the Government Accounting Office (GAO) released its assessment of 72 major acquisition programs across the four services. It found deficiencies occurring at virtually every program-management level. The study concluded that 63 percent of all programs reviewed had requirements changed after system development began, and that these same programs experienced significant cost increases. Furthermore, roughly half the programs reviewed experienced more than a 25 percent increase in the expected lines of software code since starting their respective system development. The software code finding is significant as software is prevalent in virtually every DOD system, and also because certain software programs have the ability to perform hardware functions of years past. In relation to oversight and acquisition management, GAO found that the average tenure of program managers was less than half of that called for by DOD policy, resulting in poor accountability and systemwide mismanagement. Additionally, due to manpower constraints, DOD relies heavily on contractors to perform roles that in the past have been performed by government employees. The result was about 48 percent of staffs supporting program management were from outside of the government.

The implication of these findings is that the deficiencies significantly contributed to delays and cost overruns—resulting in lost opportunity costs. Citing one instance, the report found that because of delayed deliveries and quantity reductions with the F-22A and Joint Strike Fighter, legacy systems with less capability will make up a larger proportion of the future fighter fleet for a longer period of time. Consequently, the USAF will be forced to invest billions of dollars to maintain, and eventually modernize, legacy aircraft to keep them available and capable to meet mission requirements.

Some say that the acquisition process is too complex for the USAF to effect. This ignores that the service has significant control over the process. Complicating acquisition is that while system designs may start small, by the time they reach production they have grown in size, complexity, and cost. First, **the USAF must improve its requirements definition process and establish clear guidelines that restrict introducing modifications once system development has begun**. Better definition of initial program requirements will reduce the need for change, thus shortening cycle times and better controlling cost overruns.

Further, to correct deficiencies, the USAF must adopt a strategy of standardization and an incremental approach to systems acquisition across all three domains. Standardization reduces cost and potentially shortens production times, making it possible to take greater advantage of existing technologies. Incremental development allows systems to mature while taking advantage of existing capabilities, reducing the need for modifications after system development has begun. The next generation systems are better served by development with the A model followed by B, C, and D models rather than producing the E design as the first available.

To improve requirements definition, the **USAF must increase its investment in personnel with scientific and engineering expertise**. **It**

must also stabilize and lengthen program manager tours to ensure oversight and accountability. Longevity provides familiarity which enables greater expertise in program oversight. Personnel with greater technical expertise, staying in place for longer periods of time, will lend stability to the acquisition process. Additionally, **USAF personnel—either uniformed or civilian—must perform the preponderance of program oversight, review, and continuity.** Contractors providing oversight of other contractors places stress on the system and potentially brings into question the integrity of the entire acquisition and development process.

Also, the USAF must determine if existing technologies meet mission requirements before being enticed by new programs. Mature technologies provide a cost, reliability, and time advantage over emerging systems. Former Secretary of the Air Force Edward "Pete" Aldridge once said the United States launched new, one-of-a-kind Lamborghinis into space while the Soviets launched tried-and-true Chevys. Both accomplished similar missions, but at dramatically different costs with a corresponding operational responsiveness—the Soviets could launch on demand, the United States could not.

Further, **the USAF must look to "right-tech" solutions to simplify the acquisition puzzle.** Right-tech need not mean reduced capabilities. A right-tech solution means acquiring the appropriate technological solution for a given requirement. For example, while a single satellite may be less capable than the one it replaces, networking less expensive, less complex satellites together could result in a more capable and resilient constellation at a reduced cost and in less time.

In addition to program mismanagement, consolidation and mergers within the US industrial base have also hindered the acquisition process. During the 1980s, more than 20 contractors competed for most defense contracts. Today, DOD relies principally on six main contractors, and programs have grown so complex and costly that often the six remaining contractors must team to share costs. For example, Boeing and Lockheed Martin have announced they will collaborate on the next-generation bomber. The result is reduced competition accompanied by increased cost. In a recent *Washington Post* article, Dov Zakheim, former under secretary of defense, and Ronald Kadish, former director of the Missile Defense Agency, suggested that the United States is approaching an arsenal system where government manufactures its own weaponry; a construct not unlike that used by the Soviet Union during the Cold War and the United States during the nineteenth century.

Making the issue more complex is the realization that it is extraordinarily difficult for new US corporations—beyond the big six—to compete for major weapon systems contracts. New technologies give birth to new start-up companies, but these new companies are often merged with the big six in cash buyouts or stock swaps. However, certain USAF systems are too critical to the nation's defense to allow international competition, such as computer chips and national asset satellite subsystems. **To en-**

sure these select components are both available for military use and are reliable, (*e.g.*, virus free), domestic development of state-of-the-art manufacturing designed for the most sensitive systems and subsystems is critical. However, financial markets and manufacturing process are so entwined that "buy American" often becomes impossible. For example, the recent tanker buy forced the USAF to choose between a contractor who had partnered with Europeans and another who had partnered with Asians. In fact, all of the six major prime contractors bidding on USAF contracts are global entities—as they must be to survive in the global marketplace. Thus, the nation is at a point where it can either be swept along by global acquisition or create a new acquisition strategy where it can effectively manage the process. **Therefore, acquisition protocols must be developed to enable smart global partnerships related to weapon-system acquisition that protects information sharing, software development, and hardware production**. This will not be simple, so multiple strategies will be required. Combined with a new generation of astute acquisition personnel, oversight of foreign facilities, and "no notice" quality control inspection regimes can be greatly enhanced. Ultimately, the US government must be willing to enforce stiff penalties, up to and including contract denial, to companies found in violation of quality control agreements.

Taken together, the aforementioned reforms will reduce both the acquisition cycle and overall program costs. While the United States has not lost its technological advantage, the gap between it and the world has been reduced over the past two decades. Baseball great Satchel Paige once said, "Don't look back; someone might be gaining on you." To remain uncontested, the United States must quicken the pace, or it will be forced to start looking back.

Nuclear Surety and Deterrence: USAF at a Crossroads

During the Cold War, nuclear deterrence was the nation's top priority. However, in the latest National Military Strategy, the nuclear issue has now shifted from that of deterrence to limiting the proliferation of weapons of mass destruction and downsizing nuclear assets. During 2008, many Americans were dismayed that nuclear weapons were flown cross-country, and that timing fuses were shipped abroad. This failure highlighted not just a loss of command and control, but a lack of focus on the nuclear issue, at large.

Since the end of the Cold War, the United States has struggled with nuclear deterrence as a strategic concept. Further complicating this issue is the fact that over the past 15 years the USAF has been principally engaged in conventional, operational, and tactical missions. More recently in Iraq and Afghanistan, the USAF has been engaged in air-to-ground and CAS missions. Some would say that during this time the USAF moved from the concept of *accepting no risk* in the nuclear arena to one of *risk mitigation*. As a result of this, and other actions, the USAF placed less emphasis on nuclear deterrence, a mission critical for the defense of the nation.

Numerous options exist to address the USAF's nuclear surety concerns. They range from recommendations to create an organization responsible for overseeing nuclear issues to a complete reorganization of the USAF around the nuclear mission. In reviewing past failings, it seems critical to ensuring future success that any organizational remedy be as simple as possible so that even the youngest Airman knows who has command authority for the Air Force's nuclear assets. While some may criticize the Air Force's existing command structure based on major commands and numbered air forces, it is the structure the Air Force has chosen and with which Airmen are familiar. Accordingly, the most direct organizational solution requires that **the USAF consolidate all intercontinental ballistic missiles (ICBM) and nuclear capable bombers—those so designated by treaty—under a separate major command.** When needed, tankers would "chop" to support nuclear deterrence. This major command would be responsible for organizing, training, and equipping the USAF's nuclear force, as the force provider to United States Strategic Command (USSTRATCOM). One numbered Air Force for bombers and one for ICBMs also would work to ensure a focus on nuclear weapons and improved "checklist discipline."

Additionally, **the USAF must return to no-notice inspections conducted by the NAF and higher headquarters.** Gone must be the days when nuclear weapons go "missing" or a missile alert facility or launch silo burn without personal and professional accountability. While it is true that this should not be a "one mistake Air Force," one mistake in the nuclear arena can have disastrous results for national security and international relations. Where nuclear weapons are concerned, tactical actions are linked to operational and strategic effects; far from the first order effects of the tactical mission.

Ultimately, the most recent nuclear surety problems may only be a symptom of a much larger issue within the USAF; that being a loss of strategic focus. Following the end of the Cold War, the USAF brought offensive strike fixed-wing assets under one major command—ACC. The choice before the USAF is to continue with the existing organizational structure shaped by the Cold War, or move to one based on effects, not on platforms and legacy systems. Accordingly, if the USAF is bold enough, **now is the time for an effects-based organization that would organize, train, and equip the force responsible for attacking the adversary deep in their heartland.** This approach recognizes that in the twenty-first century, offensive operations can be either kinetic or non-kinetic and be delivered through air, space, or cyberspace.

This new organization would command all nuclear assets, but would go further in bringing long-range conventional strike, offensive cyberspace, and when the time comes, space-based strike under one umbrella. This is not a proposal to "bring back SAC." While memories of Strategic Air Command (SAC) warm the hearts of many a cold warrior, a SAC construct would not meet the requirements demanded of a twenty-first century air force. **This new command would provide an advocate within the USAF for its strategic mission and be responsible for organizing, training, and equipping the USAF's strategic strike force.** For theater operations, the command would chop forces to the appropriate theater commander and "fight" under the control of the JFACC. However, for global strike, the command would provide forces to STRATCOM, commanded by the STRATCOM commander.

Organizational change alone, however, will not address much deeper problems. The larger issue is about leadership and instilling a culture where officers and senior noncommissioned officers (NCO) are not afraid to lead. Leadership has two essential elements—*the mission*, objective, or task to be accomplished, and *the people* who accomplish it. All facets of leadership must support these two basic elements. Effective leadership transforms human potential into effective performance in the present and prepares capable leaders for the future. A leader must never forget that people perform the mission.

Former chief of staff of the Air Force Gen Ronald R. Fogleman once said, "To become successful leaders, we must first learn that no matter how good the technology or how shiny the equipment, people-to-people relations get things done in our organizations. People are the assets that determine our success or failure. If you are to be a good leader, you have to cultivate your skills in the arena of personal relations." The recent failures in nuclear command and control point directly to leaders who forgot the relationship between people and the mission. The success of nuclear surety and deterrence is dependent on success at the critical junctures in leadership. These critical junctures are at the *officer to senior NCO* level and the *senior NCO to junior Airman* level.

Human behavior will drive some to attempt to find the "easy way." The Air Force core values recognize this failing and, accordingly, form the bedrock of Air Force leadership. The core values are a statement of those institutional values and principles of conduct that provide the moral framework within which military activities take place. The three fundamental and enduring values of integrity, service, and excellence require a personal focus—one that is face-to-face that directly influences human behavior and values. Successful leaders tailor their behavior toward their fellow Airman's need for motivation, achievement, and sense of belonging, recognition, self-esteem, and control over their lives. Leaders foster growth by insisting their people focus attention on the aspects of a situation or mission they control. Where Airmen assume away the importance of leadership, *compliance* with established procedures is ignored and accountability becomes questionable.

Modernizing ICBMs: America's Nuclear Deterrent

A significant part of the nation's nuclear surety issue is that it has an aging ICBM force upon which it relies for strategic defense. At its height, the United States had 1,500 land-based ICBMs. With the closing of the 564th Missile Squadron in Montana, that number is reduced to 450, most of which are single warhead ICBMs. Ongoing modernization and upgrade programs will allow the existing missiles, warheads, and command and control systems to remain operational through the 2023 timeframe, notionally to 2030, but how far beyond that remains unclear.

Nuclear deterrence remains critical to our nation's defense, particularly in light of a resurgent Russia; China adding to its nuclear force; and the posturing of nations such as North Korea and Iran. Nuclear deterrence provides the overarching umbrella to national security not offered by any other weapon system. The question with an aging fleet then becomes what does the USAF do next? Does the nation relinquish the land-based ICBM force and reply upon its strategic bombers and ballistic missile submarines for nuclear defense, or does it fund modernization of the land-based force?

America built the current ICBM command and control structures—launch control centers (LCC) and launch facilities (LF)—during the 1960s. Analysis—at an unclassified level—indicates that these underground structures have not been survivable from a direct nuclear strike for decades. Additionally, the USAF must answer the question of how much longer can they be sustained. The below-ground concrete structures that house the crews, equipment, and LFs are subject to environmental pressures, such as underground streams and shifting ground. Constant maintenance, upgrades, and modifications have kept the systems viable, but how much longer and at what cost can the USAF support an aging infrastructure while attempting to hold back time?

Further, the Minuteman III launchers will exceed their life expectancy sometime after the 2023 time span. They are viable through that timeframe only because of the current Minuteman modernization efforts. The solid propellants have been re-poured, but there is a limit to the number of times this can be done, and ultimately the missiles' engines and shells will weaken and fail. Even if extended beyond 2023, the system does not have an indefinite life span. At some point, the nation will be faced with the decision to retire or modernize the current fleet.

Support for nuclear modernization is problematic. From the public's perspective, ICBMs do not protect the nation from terrorists and have little or no role in the Global War on Terror—much less in Iraq or Afghanistan. Finally, many believe ICBMs will never be used due to the horror of nuclear weapons. Therefore, eliminating one leg of the triad would save money needed for other, more pressing requirements.

In eliminating the ICBMs, the United States would rely upon the remaining two legs, strategic bombers and ballistic missile submarines. However, the world is rapidly reaching a time when stealth and altitude—high or low—will no longer be sufficient to penetrate an adversary's airspace. Additionally, the recent Blue Ribbon review of nuclear security commissioned by Secretary of Defense Gates found that, "Without an alert commitment for 17 years . . . the bomber force has seen a dramatic atrophy of its nuclear operational and academic skills set." Thus, without returning to the Cold War strategy of bombers on alert, the preponderance of reliance and risk then moves to the submarine force, essentially requiring the United States to rely upon a single nuclear system. Presently, our submarine force is capable of roaming the seas without detection. However, if a technological breakthrough occurs that enables adversaries to determine the location of United States submarines, then the nation would eventually lose its nuclear deterrent and thereby, dramatically increase its vulnerability.

The nuclear triad has served America well for over 60 years. The rationale for its existence continues today, and will so into the future. As horrific as 9/11 was, it did not place the nation's survival at risk. Today, only a nation with strategic nuclear delivery capacity can do so. The October 1998 Defense Science Board Task Force on Nuclear Deterrence concluded that, "Significant numbers of ICBMs deny any adversary the benefit of a limited attack. Without the ICBMs, surprise attacks against a handful of bomber bases and sea-launched ballistic missile (SLBM) facilities, with plausible deniability, could drastically alter the correlation of forces." Thus, the pressures that exist to eliminate land-based ICBMs are not founded on deterrence or national defense, but rather on funding and public perceptions of nuclear utility not on external threats.

Continued reliance on a triad reduces risk and enhances national security through nuclear deterrence. Accordingly, what is required is a sustainable and affordable ICBM modernization program. The first step in creating such a system is upgrading ICBMs with a modern command and control structure. To address the aging, non-survivable LCCs, **the USAF should consider moving the equipment and crews above ground**. This move will reduce life cycle costs (ease of maintenance) while taking advantage of the latest technologies in communication. Vulnerability increases, somewhat, making this more of a soft target, but through redundancy, much of the vulnerability can be mitigated. Placing the new command and control structure within the existing aboveground missile alert facilities can also further mitigate the vulnerability, (*e.g.*, fenced and guarded). The next step in creating a sustainable force is equally important: modernization of the launch system.

The **new launch system should be based on a common vehicle**—one also used for commercial space lift or for wider military application. A missile with wide application allows research and development costs to be spread over a larger number of vehicles, while at the same time reducing maintenance costs through use of common hardware. As with command and control facilities, the USAF should consider returning the **new launchers aboveground**. Essentially the same rationale exists for moving launchers as it does for command and control: reduced life cycle costs. Again vulnerability increases somewhat, but can be mitigated through leveraging technical means, such as remote surveillance. This is not a new idea, as the USAF flirted with the concept of aboveground launchers during the

1970s and 1980s, specifically rail-garrison. This study does not propose moving to a mobile ICBM system, but this clearly is a consideration.

Affordability of a new system must also be addressed through reducing the number of ICBMs in the fleet. Additionally, reducing the number of ICBMs and accompanying warheads can be viewed as a stabilizing factor. This study does not attempt to recommend a number of ICBMs below the current 450. A new number will be a function of US national policy and should be based upon a reasoned threat analysis. Any future study determining a new ICBM force structure must take into account deterrence against rogue-nation attack and future near-peer competitors—especially as part of a triad which also encompasses bombers and subs.

Long-Range Strike: Range, Payload, Persistence—The Essence of Airpower

The value of long-range strike is as critical to the success of the future USAF mission as it was in the 1940s. Its value lies in the unique capabilities of range and payload; a bomber has proven to be the best platform to maximize these capabilities. This will remain true for the 10-to-15 year timeframe for this study. Political complexities and/or heavy enemy defenses may preclude the USAF from having adequate basing close enough to the fight to make short-range fighter aircraft viable without over-extending both the fighter and tanker force. Further, the heavy reliance on the fighter and tanker force to support one conflict creates risk in other theaters. The bomber has the range without excessive tanker support to operate from distant bases while gaining access to an area of operation. Bomber aircraft carry large payloads, giving the JFC greater flexibility in conducting operations ranging from IW to conventional campaigns, while reducing the total number of aircraft needed to achieve desired effects.

Within the air domain, the following four scenarios describe the likely operating environments in which strike operations will be needed in the 2018–23 timeframe:

1. Addressing the challenge of a near peer competitor
2. Conducting intense conventional campaigns against powers with mid-level military capabilities
3. Providing long-duration air support missions in low-threat environments
4. Providing nuclear deterrence

Future operations across these scenarios demonstrate the need for the capabilities that a penetrating bomber can provide. The issue is, will the USAF have this capability available in large enough numbers to support combatant commander requirements, especially by 2018?

The 2006 *QDR* and Congress directed the development of a new land-based penetrating long-range strike capability to be fielded by 2018. If the USAF is not fully committed to building a new bomber, the lack of commitment could create a capability gap. The B-52 fleet is aging and both the B-1 and B-2 fleets operate in limited numbers. This gap would create operational risk; a risk that becomes apparent when the scenarios outlined above are used to describe requirements.

Against opponents with anti-access capabilities—scenarios 1, 2, and potentially 4—relying mainly on short-range systems to find, fix, track, and attack the deepest and most difficult target sets follows a risky path. Resolving issues such as basing close enough to the fight, adequate tanker support, loiter capability, and payload are all complicating factors when the USAF must rely principally upon short-range aircraft. Operations requiring long-duration air support in a low-threat environment—scenario 3—range, payload, and persistence are also critical factors. Overreliance on short-range systems increases the number of strike and tanker assets needed. This increased reliance on larger numbers of short-range assets increases operational costs, while simultaneously reducing their availability to other theaters. Additionally, all four scenarios are constrained by fewer fighter aircraft numbers available to all combatant commanders, as fewer fighter aircraft will enter the inventory as the USAF's budget continues to decline and the cost of recapitalizing the fighter inventory increases.

Viewing operations in Iraq and Afghanistan, USAF leaders could judge future risk to American forces acceptable based on the rationale they are unlikely to be confronted by a capable and determined opponent across the entire spectrum of air warfare. However, using history as a guide, such assumptions incur significant operational risk. Taking the long view, it would be prudent for **the USAF to rebalance its modernization investments in the next-generation long-range strike platform to avoid a "bomber gap."** Doing so ensures the key attributes of airpower—range, payload, and persistence—are available to support the full spectrum of operations beyond the 2018–23 timeframe.

Space: The Ultimate High Ground

As a unique operating medium, space and military assets in space provide perspective, enabling awareness and responsiveness unconstrained by national boundaries. As the "ultimate high ground," space truly provides a unique arena for the conduct of USAF Global Vigilance. First, to understand space, one must understand where and what it is. Space is identified by international treaty as starting at 65,000 feet. It extends upward in all directions from the earth's surface, thus a global common or domain. Operational space exists principally of near space—defined as high altitude by some—and low and high earth orbit. While the definitions are not precise, near earth starts at 65,000 feet, low earth ranges from about 100 to 1,240 miles, and high earth orbit generally means geostationary orbit, approximately 22,240 miles up. Near earth has eddies and wind currents and is capable of sustaining high-altitude balloons. Low earth orbit is where most satellites and the space shuttle operate, and high orbit, or geostationary, is where the capability exists to position a satellite over a specific area on earth and keep it geographically stationary.

The National Aeronautics and Space Administration (NASA) and DOD are the primary organizations responsible for US space operations. Civilian-operated NASA controls manned space flight and space exploration, while within DOD, the USAF is identified as the executive agent for space and directs the preponderance of military space systems.

No one today questions the importance of space operations as an integral part of American national security. In the past 15 years, reliance on space has grown exponentially. GPS receivers are commonplace in many of today's vehicles, commercial banking is dependent upon satellite communications, and both land-based and satellite cable television receivers rely upon space-based assets. Military reliance is no less dramatic, enhancing Global Reach, and ultimately Global Power, through space-based lines of communications. From ISR to targeting, satellites provide a technological infrastructure that enables today's precision strike and superiority of the battlespace. However, the US reliance on space capabilities has also created a vulnerability—thus a likely target for potential adversaries. The recent launch of the Defense Satellite Communication System (DSCS) follow-on, Wideband Global System (WGS), is an example of this good news/bad news story. While each WGS satellite is more capable than the entire nine DSCS satellite constellation, the planned six-satellite WGS constellation increases US space vulnerabilities by placing greater reliance upon a reduced number of expensive satellites.

Assured access to space is essential to US national security. Space control, as a concept, is little different from that of air and potentially cyberspace control—and it is problematic. The United States has limited ability to either protect its space assets or deny the actions of others in space. As more nations field space systems, to include anti-satellite (ASAT) technologies, space superiority cannot be assured. The challenge for the USAF in the 2018–23 timeframe is to establish a strategy that furthers United States national interests and creates a survivable space-control architecture.

Existing vulnerabilities in space could drive a strategy that would lead the United States to place weapons in space. While this may be a viable strategy, a less expensive alternative and potentially more successful strategy for the 2018–23 timeframe exists. Using the entire spectrum of diplomatic, information, military, and economic capabilities, the **United States should develop a defense-in-depth construct for space operations**.

First, in developing a space defense-in-depth, the **USAF should take the lead in fully engaging the international community in creating a system of protocols and relationships that encourages beneficial and/or benign behavior in space**. Through economic and technical cooperation, nations become interdependent and much less likely to act against their own interests. Partnering with the international community enhances the USAF's ability to execute Global Vigilance, Reach, and Global Power. America already partners with the International Space Station, and both

Russia and China have launched satellites for United States–based corporations. Further, the United States and Europe are cooperating to avoid frequency overlap in the deployment of Galileo, a European version of GPS.

Partnering also lays the foundation for international negotiation, regulation, and governance by the rule of law, powerful concepts appreciated by our European allies. Already, the United States is party to a series of international regulations across land, sea, air, and space. International agreements limit nuclear weapons in space, but do not specifically prohibit others, such as directed energy or ASAT weapons. A new round of international agreements could call for the elimination of all weapons in space, which many nations might well find attractive. Precedent exists to regulate space activity through international negotiation and regulation. Following a successful US space-based nuclear weapons test during the early 1960s, the international arena moved to ban the use of these weapons in space.

Vulnerability in space can be further reduced by partnering with commercial industry. Currently the USAF partners in both launch and satellite production, and DOD leases commercial satellites to provide enhanced communications to the war fighter. DOD should partner with the space industry in the same fashion as has the aviation community under the Commercial Reserve Aviation Fleet construct. By providing incentives, either through direct funding or tax breaks, commercial communications and imagery companies could be persuaded to provide additional capacity for use by the military in a time of crisis. Building excess capacity into commercial satellites or warehousing spares either in orbit or on earth, the United States can purchase capability at a fraction of the cost of a dedicated DOD space system.

Pres. Ronald Reagan once said, "Trust, but verify." In space this is problematic, for without situational awareness it is difficult to do either. Many of our land-based radar systems were originally designed for ICBM early warning, not objects in low earth or geosynchronous orbit. Currently, an adversary could alter a satellite's orbit by a few degrees, and it may take days or weeks to reacquire. Additionally, microsatellites are a reality, and the United States has little or no ability to track objects that small. The USAF has taken positive steps to correct the deficiency with the launch of the first space-based surveillance system (SBSS), Pathfinder, in an attempt to improve space situational awareness of objects in geosynchronous orbit. However, the **USAF must fully fund SBSS, develop, and field a space constellation designed to detect objects in low earth orbit, and integrate space and ground systems into a coherent architecture**. Only then will the United States be capable of providing comprehensive space situational awareness, an essential element for ensuring true space superiority.

Another essential element to space control is access to the domain. The USAF has a rich history of being involved in the nation's race to space. It does not, however, have a record of responsive launch. Special handling requirements for lift vehicles and satellites require months of planning for any on-time launch. **Space systems must become more responsive and**

less vulnerable at the same time. This can best be accomplished through the concept known as Operationally Responsive Space (ORS). Under Secretary of the Air Force Peter B. Teets, in testimony before Congress, 25 February 2004, defined ORS as a means "to create a more responsive, reliable, and affordable lift family capable of fulfilling both current and future launch requirements, and the corresponding responsive and affordable satellites." Under this construct, ORS is further defined as developing space assets that are responsive to the war fighter.

The primary lifters for today's USAF, known as the *evolved expendable launch vehicle* (EELV), are Boeing's Delta IV family and Lockheed Martin's Atlas V family, which will continue to be the primary medium and heavy lifters well beyond the 2023 timeframe. Becoming operational in 2002 and at about $100 million per vehicle, the EELV was designed to standardize and improve space launch operability, reduce the government's traditional involvement in launch processing, and save a projected $6 billion in launch cost between 2002 and 2020.

In 2006, a congressionally mandated national security space launch requirements panel addressing DOD lift concluded that, "ample evidence suggests that these rockets [Delta IV and Atlas V] can meet the NSS [National Security Strategy] launch needs of the United States through 2020 (the end of the [congressional] study period), barring the emergence of payload requirements that exceed their design lift capability." The report noted, however, that the two launch families were "largely uncompetitive in today's commercial market," and that because ORS concepts were in the formative stages, "it was premature to specify launcher requirements." The USAF objective must be lower cost, with responsiveness marked by days and weeks rather than months and years. **Smaller, less expensive lifters and satellites must become commonplace**.

To further mitigate vulnerability in space, the United States must reduce its reliance on any single satellite or limited grouping (*i.e.*, limited constellation) of satellites. This can be accomplished by having spares on orbit or the ability to rapidly replace lost assets. If an adversary negates the use of one or more satellites, through responsive launch and the ability to rapidly reconstitute the constellation, the United States negates this vulnerability. Further, by networking less complex satellites together, as is done today with computers, operational capability can be greatly enhanced. The nation will always require large and correspondingly more expensive satellites, especially in geostationary orbit. However, what is needed is a mix of both systems to increase capability and simultaneously reduce vulnerability.

The USAF has recognized the requirement for ORS and is moving in the right direction with the Minotaur IV, a modified Peacekeeper ICBM that reduces cost with a smaller lifter for smaller payloads. However, this approach is not sustainable despite the inventory of available unused Peacekeepers. Partnering with industry, as the USAF is doing with the commercial company *SpaceX*, will be necessary to reach the goal set by Secretary

Teets. Commercial companies are currently testing low cost lift in the range of $7 million per launch, producing a cost-to-weight lift ratio even less than that of the $30 million per Minotaur.

To date, what is missing is a commitment to ORS beyond a test program. **The USAF must step up to this challenge or it will remain mired in the construct of vulnerable, expensive, one-of-a-kind systems. Additionally, the time is right for a second *Space Launch Modernization Study* similar to the one in 1995 that produced the requirements for EELV.** The new study should identify the requirements for ORS, bringing divergent USAF lift and satellite programs together, and focus on light-to-medium vice medium-to-heavy lift. Only when requirements are established will ORS move from a test program to an operational concept.

Range modernization is another area of concern for the nation to ensure space control. The USAF has been on a steady program to improve facilities at Patrick AFB and Vandenberg AFB over the past 20+ years. Both ranges are, however, decades old and this alone necessitates increased investment that many competitor nations do not face. Far worse is that the US ranges are at locations that require 25 percent greater lift capability over those located on or near the equator. The United States will continue to need its CONUS launch ranges, particularly for sensitive payloads, but it can move to a lower cost *via* more capable alternatives for routine lift. **Through partnering, the United States could gain access to more economical lift**. Nations such as Brazil offer the potential for modern LFs at reduced costs, where less thrust is required to place an object in orbit. Further, combining **on-site fabrication facilities** for satellites and lifters could reduce cost and enhance responsiveness.

The USAF has one significant gap in its exploitation of space, that being near space—known to some as high altitude. Near space is the region sandwiched between an altitude of about 12 miles and 62 miles, loosely defined as the lower limit of space. In this region the air is too thin to support flight by most aircraft, and gravity is too strong for a satellite to sustain orbits. Near space provides an arena to further Global Vigilance through theaterwide surveillance and Global Reach *via* new technologies that deliver alternative and potentially more flexible lines-of-communications systems for theater battlefields.

While chief of staff of the Air Force, Gen John Jumper stated that near space vehicles would be an inexpensive substitute "for a low orbiting satellite constellation." The Defense Advanced Research Projects Agency (DARPA) and AFSPC have successfully tested vehicles at this altitude, DARPA using a "flying wing," and AFSPC with high-altitude balloons. The goal of exploiting this arena is to provide greater capability for the war fighter at a substantially reduced cost. The expectation is that if the USAF could build a near-space vehicle to hover over a point—at an altitude of about 23 miles—it could remain on station for months, far longer than an unmanned aerial system and a period approaching the mission duration of some satellites, at a greatly reduced cost.

Recently, operational enthusiasm for near space has waned. The USAF has shifted the responsibility for near space from AFSPC to ACC. Despite lack of emphasis, this arena remains inherently a USAF medium. **Leadership must commit now to the use of near space to ensure operational control of a cost-effective medium for capitalization in the 2018–23 timeframe.** Ultimately, near space is awaiting exploitation across its entire spectrum with the potential for great return on limited investment.

Cyberspace: Boundless Opportunity and Significant Vulnerabilities

During the earliest days of the computer revolution, cyberspace was largely viewed as a benign, undefined region whose main use was for electronic mail and a playground for pimply-faced teenage boys and twenty-somethings living in the basement of their parents' house. By contrast, rapid growth in the popularity of the personal computer and the increased availability of the Internet has generated a new warfighting domain: cyberspace.

Today, cyberspace binds the international community together, empowering governments and individuals in ways unimaginable only a few years earlier. Cybers+pace permeates nearly every aspect of our daily lives to the point where society depends on the use of information technology networks. However, as with space, our reliance on cyberspace has turned a technological advantage into a vulnerability that adversaries seek to exploit. The challenge, then, is how best to make use of the medium, while simultaneously protecting United States national interests from attack *via* this new domain.

Cyberspace, a term coined by author William Gibson in his 1982 work *Burning Chrome*, was defined by the author as "a graphic representation of data abstracted from banks of every computer in the human system." From this fictional beginning, the world would grasp the concept of a real, separate, distinct, and identifiable realm. In 2008, DOD defined *cyberspace* as "a global domain within the information environment consisting of the interdependent network of information technology infrastructures, including the Internet, telecommunications networks, computer systems, and embedded processors and controllers." (Dep SecDef letter, 12 May 2008)

The DOD definition, according to Lt Col David Fahrenkrug, former Eighth Air Force chief strategist, means that cyberspace is a very real, warfighting domain. DOD senior scientist, Dr. Kamal Jabbour, has described *cyberspace* in military terms as "a domain in which signals hold at risk intelligent systems." As such, cyberspace enables Global Reach—not to mention Global Power and Global Vigilance—through a modern line of communication that connects all other domains *via* physical infrastructure and the cognitive processes that use the data that is stored, modified, or exchanged.

Cyberspace's distinct characteristics immediately identify it as unique from the other warfighting domains of land, sea, air, and space. The characteristics are the low cost of entry—anyone with a computer and an Internet connection can launch attacks against global United States interests, anonymity of one's action, and uncertain governmental jurisdiction due to international cross-border implications. At the same time, cyberspace provides the ability to deliver precise effects that enable technologies and tactics to operate simultaneously across a broad range of targets, unconstrained by location or time.

Cyberspace is also distinct from the information that may be resident in or transferred through it. If not properly understood, in an attempt to protect information, the USAF might err in focusing on the content rather than on the domain itself. A maritime analogy would be the equivalent of guarding the goods on board a ship rather than patrolling the shipping lanes. Accordingly, the **USAF must not focus solely on protecting its databases, but rather it must also protect its networks and the functioning of electronic devices to enable cyberspace control.** Threats will come from traditional and nontraditional sources, including hackers in search of fame or personal gratification, criminals seeking profit, terrorists looking for ideological gain, and nation-states in quest of political and/or military advantage.

Establishing control over cyberspace does not mean having exclusive use of the domain, nor does it mean that the USAF is interested in defending all of it. Instead, as with air superiority, control of cyberspace means securing access to certain portions of the domain as needed to conduct desired activities across all domains. **Accordingly, the USAF must be prepared to conduct warfare in cyberspace in order to secure the domain at the time and place of its choosing.**

Before the USAF can function effectively in cyberspace, it must first resolve the issue of jurisdiction, thereby determining the difference between a crime and an act of war. This is not as simple as it might sound. The *Posse Comitatus* Act generally prohibits DOD personnel from taking law enforcement actions against United States citizens, which prompts certain questions. When an attack occurs, is it from inside or outside of the United States? Is it by a United States citizen or a foreign national? Is it by a group of unaffiliated private citizens or is it state sponsored? Where does the Federal Bureau of Investigation's (FBI) or National Security Agency's (NSA) authority begin in relation to that of the USAF? The answer to these questions will drive jurisdictional decisions and provide the authority for the USAF to take offensive or counteroffensive cyberactions.

The USAF must take an aggressive approach to determining jurisdictional responsibility. Accordingly, **the USAF should call for an interagency commission that will resolve the issues of jurisdictional authority and, if needed, advocate for Congress to revise Title 10 laws and policies**. Without resolving this issue, it will become increasingly difficult to ensure operational control of the domain required to defend the nation.

Understanding cyberspace as a war-fighting domain requires focusing on the medium as it relates to military operations and national security. The ability of an individual or state to "hack the system," or "SPAM the network," resulting in denial of service or corrupted databases, will only increase over the next 10 to 15 years. Accordingly, **while offense offers a distinct advantage for airpower, in cyberspace deterrence and defense must become coequal propositions**. When attacks can come from all directions at any time—and at the speed of subatomic particles—defense becomes paramount. Two examples follow where the USAF must take a "deter and defend" approach.

First, the open systems DOD operates invite the potential for "virus sharing" among open and closed systems alike. When we bring thumb drives from home after working on the latest briefing or rip a compact disc from a music-sharing Web site so that we can listen to it on our computer, we invite viruses into what we otherwise think of as a protected system. **The only solution to this issue is to reclaim the Internet. That does not mean drive civilians off the World Wide Web; that genie is long out of the bottle. What is required, though, is a truly closed network, one that does not allow interaction with its civilian counterpart in any capacity or other open systems across the DOD.** By moving to a closed network, the potential for external contamination is dramatically reduced. Further, a closed network provides increased defense from external attack, making it more difficult to determine an entry point for hacking or spamming.

Additionally, the USAF must defend its critical databases from attack. The USAF no longer has the luxury to pretend its bases are like feudal kingdoms and castles. Computer firewalls, much like moats and castle walls during the Middle Ages, confine the populace "inside" and restrict their actions. Also, they provide only limited protection while presenting an interesting target for the curious. Like the Maginot Line of World War II, firewalls provide a feeling of false security even as the enemy uses maneuver tactics to circle around the exposed flanks. Rather than seeing bases as individual entities working in cyberspace, the USAF must view cyberspace holistically, developing organizations and tactics to defend regardless of location, while retaining freedom of action for our forces.

Anticipating successful attacks against firewalls, **the USAF must develop resiliency in its systems, where a layered defense in depth reacts to threats and sets in motion procedures for postattack recovery.** Much like the construction of battleships of the early twentieth century, the ability to sustain a direct hit and continue fighting will be paramount. **This will mean developing and fielding self-diagnosing and self-healing systems with adequate redundant capacity for survivability.**

Conducting offensive and defensive cyberspace operations will require an increased degree of automation. Cyberoperations occur in the compressed time of milliseconds. This pace demands automating the defensive measures of threat detection, classification, course-of-action selection, and response. Intelligent systems must be able to react rapidly, instantaneously differentiating between inquiry and attack, and communicating our resolve or offensive operational intentions to deter the attacker. Further, databases must employ stealth methodologies where, for example, modulating chip technology enables them to hide, morph, and masquerade as effectively as any attacking agent.

If deterrence fails, the USAF must have the ability to conduct counter-cyberspace operations across the entire grid to disable and defeat the attacker's capabilities, "returning fire" when necessary based on established rules of engagement. By thwarting attacks, the USAF deters future actions and removes the aggressor's motivation, be it an individual or nation-state. Without a return on the investment of time and effort, attacks are thwarted, and cyberspace superiority, like air superiority, becomes achievable.

To respond effectively, the USAF must first know that an attack is taking place, thus counter-cyber operations will require generating and maintaining cyberspace situational awareness (CSA). Global Vigilance begins with knowing that attacks against US systems will likely begin by probing and querying to determine weaknesses and our likely response. Defeating a cyberattack will then necessitate internal and external early warning systems, much like the ICBM launch-detection network. Systems residing outside the firewall will identify anomalies, rapidly analyzing an attacker's "forensic fingerprints" to predict future behavior, and communicate viable options through reach-back capabilities to avoid the threat.

Gen Henry H. "Hap" Arnold said, "Offense is the essence of airpower." This holds true for cyberspace as well. Operating in cyberspace provides the opportunity for Global Vigilance, Reach, and Power, instantaneously using new ways and means to achieve traditional types of effects. Offensive cyberspace operations provide the ability to exert kinetic-like effects using electrons. Demonstrating one such capability, the Department of Homeland Security conducted a simulated attack against a power plant. Using computer network attack capabilities, they gained access to a supervisory control and data-acquisition device and sent an electrical generator out of control, causing it to self-destruct. This same type of effect can be exerted on air-defense or computer-network systems to temporarily blind America's adversaries.

Technological solutions are only part of the cyberspace challenge. A new domain requires developing a new corps of professionals capable of waging cyberwarfare. While the enlisted force will have a role, officers must take the lead as they have done in air and space operations. Thus, the USAF must identify a select group of officers whose mission will be to deliver cyberspace options for the defense of the nation and its global interests. The USAF has proven it can turn history majors into space operators and pilots, but following this model with cyberspace may be problematic. Arguably, cyberspace is the most technically demanding on operators of the three domains and will, at least initially, require professionals with specialized degrees in such fields as physics, electrical engineering, computer science, and mathematics. **Recruiting these "cyberwarriors" will require a "cradle to grave" approach.** To begin, the USAF must engage the right people. This means offering science and engineering scholarships to students who come from the nation's elite civilian institutions, such as the Massachusetts Institute of Technology, Georgia Tech, and Cal Tech.

As an operational career field, cyberwarriors must be part of a highly structured professional-development program. A specialized training curriculum, similar to UPT or Undergraduate Space and Missile Training (USMT) must exist. This undergraduate cyberspace training (UCT) school will provide the needed finishing military education required for newly minted second lieutenants. It must be as exacting as UPT, where the USAF anticipates a substantial number of washouts as they progress through a highly demanding and rigorous program. The new school would focus on the fundamentals of cyberspace control and operations, furthering a student's understanding of cyberspace superiority and cyber counteroperations.

Following UCT completion, cyberprofessionals should anticipate a career-long continuing education process. A schoolhouse similar to the National Space Studies Institute in Colorado Springs is required where 100-, 200-, and 300-level courses are available for officers as they progress in rank and responsibility. **The courses at the NCSI would provide an increased understanding of cyberoperations appropriate for success at the advanced ranks. Attendance at NCSI, combined with annual professional development requirements, will help to ensure a career force current in the latest technologies.**

The USAF must also ensure adequate pay, attendance at the right schools—PME and weapons school—and promotion. Pay must be adequate to avoid the USAF becoming the postgraduate training ground for industry, thus incentive cyberpay may need to become a reality. The incentive pay need not attempt to offset the difference between an officer's pay and that of industry, but as with flight pay, it must be adequate to reward those who choose to serve their nation, but at a reduced wage.

In developing a career path for cyberwarriors, the USAF should look at its success in air and space. Major commands, such as ACC, AMC, and AFSPC, ensure a nurturing career field for air and space professionals, and provide a pyramid rank structure where exceptional young officers have the opportunity for command and promotion with no glass ceiling that may limit their abilities. **The USAF must find a home and advocate for future cyberwarriors, one equal to that of air and space.** It will take time for cyber colonels and generals to fill command and senior staff positions, just as it did with AFSPC. However, if managed properly, in less than a generation, cyberspace will have one of its own seated at the table as an equal partner with air and space operators, wearing the weapons school patch and four stars.

Cyberspace also provides an excellent opportunity for Total Force application. Following the precedence established in air and space, the USAF should incorporate the Guard and Reserve into the cyberspace mission. **The first step is ensuring billets exist within the ANG and AFR for cyberwarriors who separate from active duty.** This ensures the USAF benefits from their investment in developing this new group of professionals, while allowing these cyberwarriors the opportunity to continue serving their nation in uniform. **Further, the USAF must develop a construct to incorporate Guard and Reserve into the "fight." Either through stand-alone units or as a part of a blended force, Guard and Reserve personnel can play a significant role in defending the nation *via* cyberwarfare.** Who would not want the civilian program manager in a Silicon Valley dot-com providing that same expertise to the nation as a part of the uniformed Total Force?

The USAF must also develop a "cyberspace red team" whose mission is to probe DOD networks and provide input for offensive, defensive, and counteroffensive strategy. The Cyber Red Team must be comprised of military and industry cyberprofessionals constantly challenging USAF systems and dedicated to improving cyberoperations by providing feedback that could be implemented to improve both defensive and offensive operations. Working closely with Air University's Curtis E. LeMay Center for Doctrine Development and Education at Maxwell AFB, the strategies developed and weaknesses discovered by the team would drive cyberspace doctrine at a pace that would rival that of the advance of early airpower doctrine.

No operator should ever have to ask, "Will my weapon work?" However, cyberspace warfare creates the possibility for Joint Direct Attack Munitions to function perfectly on the practice range at Nellis AFB, while failing against enemy targets. By embedding positional coordinates on a computer chip, a weapon could be rendered useless when employed against a specific latitude and longitude. Quality control then becomes paramount in the acquisition and manufacturing process. While requirements may be placed upon contractors, what controls exist on the subcontractors in India, China, and Indonesia building the computer chips that support the operating systems for our weapons? A real-life example follows:

> A computer virus recently discovered on digital photo frames has been identified as a powerful new Trojan Horse from China that collects passwords for online games, and its designers might have larger targets in mind. "It is a nasty worm that has a great deal of intelligence," said Brian Grayek, who heads product development at Computer Associates, a security vendor that analyzed the Trojan Horse. The virus recognizes and blocks antivirus protection from more than 100 security vendors, as well as the security and firewall built into Microsoft Windows. It downloads files from remote locations and hides files, which it names randomly, on any PC it infects, making itself very difficult to remove. It spreads by hiding itself on photo frames and any other portable storage device that happens to be plugged into an infected PC. The authors of the new Trojan Horse are well-funded professionals whose "malware" has "specific designs to capture something and not leave traces."*

Accordingly, some weapon systems and subsystems are too important to rely on the lowest bidder. The answer is not increased regulation, but rather **revitalizing and protecting the US industrial base upon which**

---

*Deborah Gage, "Virus from China the Gift that Keeps on Giving," *San Francisco Chronicle*, 15 February 2008, http://www.sfgate.com/cgi-bin/article.cgi?f=/c/a/2008/02/15/BU47V0VOH.DTL&type=business.

production of critical systems and software depends. The USAF must encourage software development and the manufacturing of state-of-the-art computer chips and subsystems inside US borders by US industry. While American production does not guarantee "virus free" components, it does enable better quality control while inspiring greater confidence.

Cyberspace acquisition will also embrace a new list of contractors. This list will include some of the major aerospace and computer industries, but more often will consist of smaller software and hardware firms whose innovative nature will drive technological development. **To ensure a viable acquisition strategy, the USAF should develop an industry council where senior military and industry representatives convene to establish requirements and propose technological solutions.** This council should meet on a regular basis, often enough to encourage an open dialogue between the USAF and the innovative elements within private industry. Gathering senior leaders together from industry and the military ensures being better informed on current technological breakthroughs, thus allowing the USAF to shorten the acquisition timeline for fielding state-of-the-art technologies.

Further, the USAF must partner with the private sector—universities and commercial industry—to properly leverage American expertise. Despite the extent of expertise within the USAF, the private sector will continue to define the edges of the technological and cyberspace revolution. The job of the new core of cyberprofessionals will be not only to innovate, but also to translate civilian expertise into national defense capabilities. Thus, partnering will be necessary to provide the USAF with state-of-the-art hardware and software, as well as offensive and counter cyberspace operations capabilities. Also, the private sector can provide "on call" capacity not resident in the USAF. Once again, space operations provide a viable model. When a satellite fails to function properly, an "anomaly resolution team" convenes. This team, composed of the military, DOD civilians and contractors, makes recommendations on possible solutions. Following this example, a cyberteam could bring similar experts together to develop courses of action against an ongoing network attack, or to determine effective counter-cyber options. This approach makes expertise available when needed and ultimately reduces operating costs in an era of static or declining budgets.

Just as cyberspace brings enormous challenges, it also promises great reward. In this domain, tremendous power can be exerted through the flow of information—packed as electrons. Reach is immediate and global, but requires constant vigilance against attack. Everyone understands the inconvenience when e-mail or the Internet goes down; however, we have ways to cope. From making a phone call, to using a fax machine, to going to the library to look up a reference, we find work-a-rounds to deal with the loss of connectivity. What the USAF cannot compensate for as readily is the destruction or theft of classified information, or hidden code in weapon systems that activate when least expected. Success will require a new core of cyberprofessionals: men and women who will not only ensure the USAF maintains its technological lead in defending the nation, but will also chart the future in this newest frontier. Their success or failure depends on the actions that the USAF takes today. These actions must lay a strong and supportive foundation for the future.

Chapter 6

Summary of Recommendations

Current Fight

1. Irregular Warfare: Winning the Long War

a. Change objective from fighting IW to that of enabling a partner to fight IW

b. Have full capability to directly engage the enemy—to execute/perform IW engagements as a key component of the national effort

c. Create within the partner nation the requisite skill sets and disciplines in air, space, and cyberspace that enable that partner to realize its national goals

d. The USAF should shape its force for a winning strategy in the Long War to enable Airmen to *assess*, organize, train, equip, *assist*, and *advise* foreign air forces for success against irregular adversaries

e. Develop a right-tech aircraft that meets both immediate IW requirements and provides the capability for partnering with host air services where a platform can be assimilated by host nation forces, creating capacity where none may have existed before

f. Through force development, prepare every specialty, though perhaps not every Airman, for IW

g. Dual qualify pilots in their primary aircraft and a right-tech IW aircraft

h. Every UAS pilot operator should be mission qualified in the right-tech IW aircraft

2. Air Mobility

a. Consider extending the line for additional strategic airlift platforms beyond those currently planned

b. Explore the potential to employ modern airships to bridge the gap between rapid fixed-wing airlift and fast sealift platforms

c. Intra-theater airlift assets of the future must be capable of operating from short, austere, unimproved airstrips while maximizing small unit lift requirements to facilitate their usefulness throughout a distributed combat environment

3. **Strategic Communication: Spreading the Word**

   a. Develop IO to the same extent as every other weapon system in the Combined Air Operations Center

   b. As a long-term goal for force development, the USAF must ensure that IO experts are better developed and sustained

4. **Culture and Language in the Expeditionary Air Force**

   a. Include a sustainment package for force development planning for languages and culture to ensure future resource proficiency and availability

   b. Determine hard requirements for languages and then plan accordingly

   c. Respond to *DoD Language Transformation Roadmap* with a comprehensive, sustainable language "flight plan," encompassing all disciplines and emphasizing a long-term commitment to language education and training

   d. To produce enough linguists in a variety of languages, each newly minted intelligence officer should be assigned a language from the DOD's Strategic Language List—it would be their professional responsibility to maintain proficiency in a language for their entire career

   e. Consider moving the service program manager for language from HAF/A-1 to HAF/A-2

5. **Distributed Planning: The Key to Centralized Control**

   a. Redefine its operational doctrine to capture the concept of distributed planning below the AOC level

   b. Place experts with the appropriate planning tools at the locations where operational plans are born and refined

   c. Develop a distributed planning structure to meet the needs of current and future operating environments

   d. Develop an organizational construct that adequately provides a JTF staff with the needed planning expertise

   e. Identify and track individuals with appropriate expertise to serve in the ACCE for JTF–level planning integration and track them in the personnel system

   f. Provide qualification and currency training so Airmen are prepared to support a JTF when one is established

   g. Place personnel with planning expertise at the brigade level and below

h. Establish a new unit that is organized, trained, and equipped to conduct distributed planning

i. Invest in resources necessary to grow the required planning expertise

6. Unmanned Aerial Systems (UAS): An Air Force in Transition

a. Develop a separate UAS operational career field, with a separate qualifying weapons school

b. USAF should take the lead in determining common language and procedures between the Justice Department, the FAA, and DOD to enable widest possible use of UASs at home

c. Determine UAS operational missions—such as reconnaissance, strike, electronic warfare—in theater, accountable to the JFACC and then defined doctrinally in the USAF and Joint communities afterward as a statement of best practice

7. Intelligence Reform: The Secondary Effects of Merging Intelligence with Surveillance and Reconnaissance

a. Emphasize operational analysis with the sensor-to-shooter chain

b. Create a separate targeting career field for AF and Joint targeting training

c. Reestablish an indigenous HUMINT capability

d. Resource OSI to accomplish HUMINT mission in CENTCOM AOR

e. Re-energize unit intelligence assignment as a building block for understanding support to the warfighter

f. Do not return ISR to the A3 community, *i.e.*, USAF/A2/3, but common functions below the A2 and A3 should be linked together to foster effective management and oversight

8. The Total Force in Transition

a. To return the balance to the "citizen-Airman" construct, the Guard's primary mission must become homeland defense

b. The ANG must place a priority on homeland defense planning in conjunction with state homeland defense authorities and their Army National Guard counterparts

c. Modify ANG medical units DOC statements designating them to perform disaster relief missions

d. Homeland defense DOCs for rescue and airlift should be added as secondary DOCs for both the ANG and AFRC units

e. ANG should codify its domestic counter-air role within a DOC statement for selected units

f. ANG will need to identify personnel with gerontology expertise, either as a primary or secondary specialty to be prepared for post–disaster relief

g. The three AFRC NAFs must assume a near mirror-image warfighting capability like their active duty counterparts, or be downsized or eliminated

h. As current equipment is programmed for retirement in out-years, selected ANG units be re-equipped with dual capable platforms and be assigned DOCs for multiple missions areas

i. Air Force IRR should be energized as the USAF's strategic reserve and resources allocated for its management and utilization

9. Air Base Disaster Planning

a. Conduct a natural disaster base vulnerability study of all active duty Guard and Reserve AF bases

b. Develop clear plans for reconstituting those bases that have a high probability of being evacuated during natural disasters

10. Recasting Global Vigilance, Global Reach, Global Power

a. Enablers for achieving strategic ends is strategic partnering

b. Global Vigilance: Situational awareness—required for understanding both the necessity for action and the character of the effects necessary to achieve a revised condition or end state

c. Global Reach: Operational access—connectivity to the objective through air, space, and cyberspace

d. Global Power: Ability to create and sustain effects through air, space, and cyberspace

Future Fight

11. Acquisition Reform: A Requirement for Ensuring Future Viability

a. Adopt a strategy of standardization and an incremental approach to systems acquisition across all three domains

b. Shorten the acquisition cycle to decrease cost and retain the current technological lead the USAF holds over other air services

c. Stabilize and lengthen tours for acquisition managers

d. Develop state-of-the-art manufacturing for select systems within the United States

e. Develop protocols to enable smart global partnerships related to weapon system acquisition that protects information sharing, software development, and hardware production

f. Improve the requirements definition process

g. USAF personnel, uniformed or civilians, must perform the preponderance of program oversight, review, and continuity

h. Establish clear guidelines that restrict making system modifications once development has begun

i. Increase investment in personnel with scientific and engineering expertise

j. Review existing technologies for application before developing new technologies

k. Look to a right-tech solution for all USAF systems

12. Nuclear Surety and Deterrence: USAF at a Crossroads

a. Near term: Consolidate all ICBMs and nuclear capable bombers under a designated major command

b. Return to no-notice inspections conducted by the NAF and higher headquarters

c. Establish an effects–based strategically focused organization that would organize, train, and equip the force responsible for deep attack, either kinetic or non-kinetic

13. ICBM and Bomber Modernization

a. Continued reliance on a triad through a sustainable and affordable ICBM modernization program

b. To address the aging, non-survivable LCCs and LFs, the USAF should consider moving the equipment and crews above ground.

c. Move to a common launcher for ICBMs

d. Rebalance modernization investments in the next generation long-range strike platform to avoid a bomber gap

14. Space: The Ultimate High Ground

a. Develop space defense-in-depth to decrease vulnerabilities

b. Partner with the international community to develop treaties and agreements to further peaceful space exploitation

c. Partner with commercial industry to provide capabilities on commercial satellites that could be used by DOD in a time of need

d. Develop space situational awareness with a fully funded Space Based Surveillance System (SBSS) for geostationary orbit and a new system for low earth orbit

e. Fully commit to operationally responsive space

f. Smaller, less expensive lifters and satellites must become commonplace

g. To further mitigate vulnerability in space, the United States must reduce its reliance on any single satellite or limited satellite constellation

h. Commission a *Space Launch Modernization Study* for light to medium lift requirements

i. Partner internationally to modernize launch ranges and system processing

j. Commit to exploiting near space/high altitude

15. Cyberspace: Boundless Opportunity and Significant Vulnerabilities

a. USAF must not focus solely on protecting its databases, but rather it must also protect its networks and the functioning of electronic devices to enable cyberspace control

b. USAF must be prepared to conduct warfare in cyberspace to secure the domain at the time and place of its choosing

c. Call for an interagency commission to resolve the issues of jurisdictional authority and, if needed, author legislation for Title 10 revision

d. While offense offers a distinct advantage for airpower, deterrence and defense must become co-equal propositions in cyberspace

e. Reclaim the Internet, moving to a closed network, one that does not allow interaction with its civilian counterpart in any capacity or other open systems across the DOD

f. View cyberspace holistically, developing organizations and tactics to defend regardless of location, while retaining freedom of action for our forces

g. Develop system resiliency, with a layered defense in depth, that reacts to threats and sets in motion procedures for post-attack recovery

h. Develop and field self-diagnosing and self-healing systems with adequate redundant capacity for survivability

i. Develop systems to support real-time cyberspace situational awareness (CSA)

j. Develop a new corps of professionals capable of waging cyber warfare

k. Develop a structured professional development curriculum with a UCT school that provides the needed military education required for newly minted second lieutenants

l. Create a National Cyberspace Studies Institute (NCSI) that provides an increased understanding of cyber operations appropriate for success at the advanced ranks

m. Ensure adequate pay, attendance at the right schools—PME and weapons school—and promotion

n. Find a home/advocate for future cyber-warriors, one equal to that of air/space

o. Identify Guard and Reserve billets for cyber-professionals who separate from active duty

p. Develop a construct to incorporate Guard and Reserve into the "fight," either through stand-alone units or as a part of active duty units

q. Develop a Cyberspace Red Team to probe DOD networks and provide input for offensive, defensive, and counter offensive strategy.

r. Develop an industry council where senior military and industry representatives convene to establish requirements and propose technological solutions

s. Partner with the private sector—universities and commercial industry—to properly leverage American expertise

Appendix A

Title 10

The National Security Act of 1947 created a "national military establishment" (NME) which was a dramatic change from before. The NME clearly lacked the hierarchy of controls associated with today's Department of Defense (DOD). DOD was created in 1949 during a series of changes to the original act.

The National Security Act of 1947 established the United States Air Force (USAF) as a separate service publishing its purpose as follows:

> In general the United States Air Force shall include aviation forces both combat and service not otherwise assigned. It shall be organized, trained, and equipped primarily for prompt and sustained offensive and defensive air operations. The Air Force shall be responsible for the preparation of the air forces necessary for the effective prosecution of war except as otherwise assigned and, in accordance with integrated joint mobilization plans, for the expansion of the peacetime components of the Air Force to meet the needs of war. (61 Stat. 502)

Title 10 governs the nation's armed forces and further defines the purpose of the USAF:

> It is the intent of Congress to provide an Air Force that is capable, in conjunction with the other armed forces, of preserving the peace and security, and providing for the defense, of the United States, the Commonwealths and possessions, and any areas occupied by the United States; supporting the national policies; implementing the national objectives; and overcoming any nations responsible for aggressive acts that imperil the peace and security of the United States. (10 USC 8062)

Title 10 then repeats verbatim the language noted above from the 1947 National Security Act regarding the Air Force and continues with the following:

> The Air Force consists of the Regular Air Force, the Air National Guard of the United States, the Air National Guard while in the service of the United States, and the Air Force Reserve; all persons appointed or enlisted in, or conscripted into, the Air Force without component; and all Air Force units and other Air Force organizations, with their installations and supporting and auxiliary combat, training, administrative, and logistic elements; and all members of the Air Force, including those not assigned to units; necessary to form the basis for a complete and immediate mobilization for the national defense in the event of a national emergency. (10 USC 8062)

Title 10 also includes language regarding ceiling authorizations for the Air Force:

> The authorized strength of the Air Force is 70 Regular Air Force groups and such separate Regular Air Force squadrons, reserve groups, and supporting and auxiliary regular and reserve units as required. There are authorized for the Air Force 24,000 serviceable aircraft or 225,000 airframe tons of serviceable aircraft, whichever the Secretary of the Air Force considers appropriate to carry out this section.

Appendix B

Funding Lines

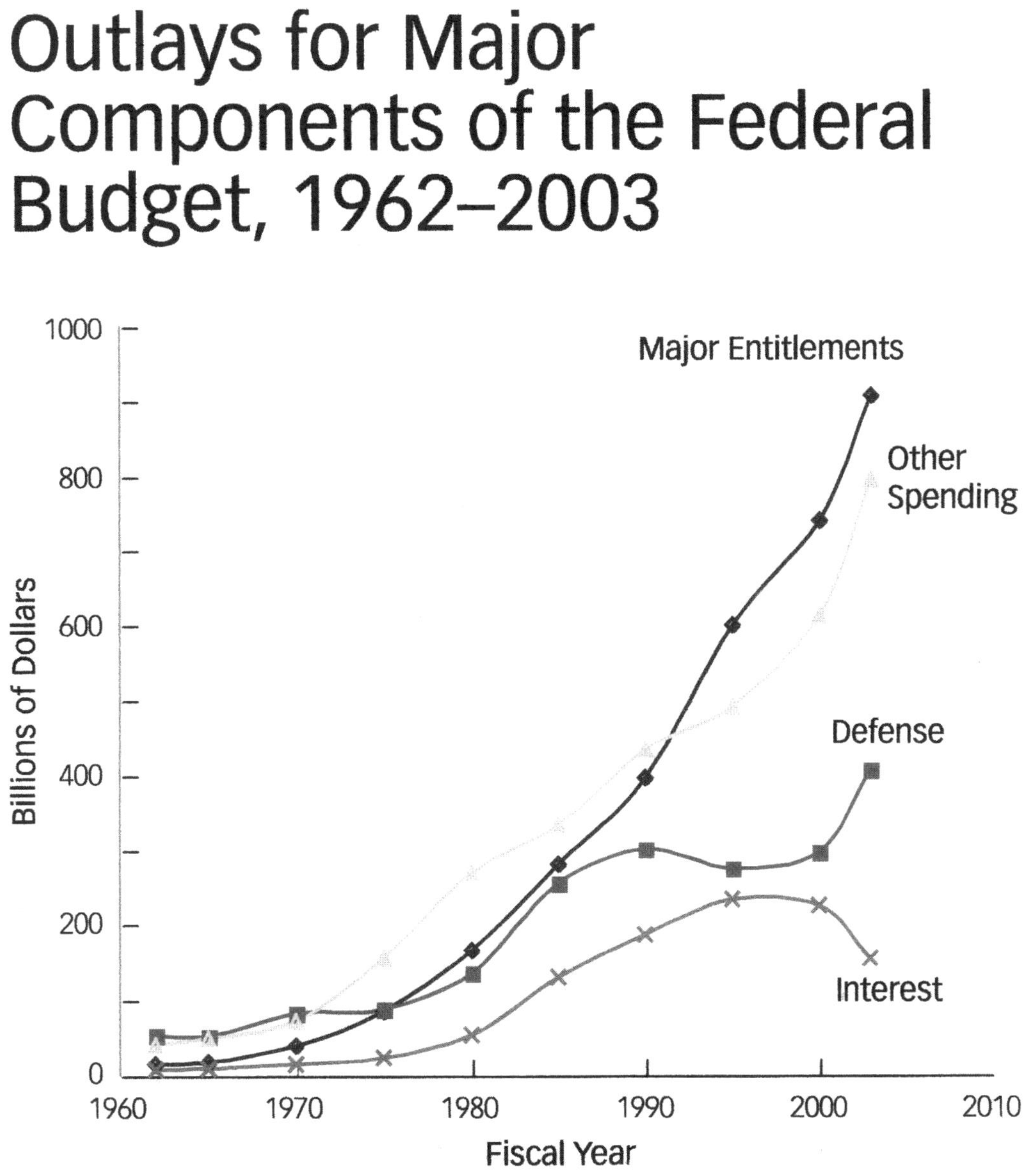

Fig. B.1. Outlays for Major Components of the Federal Budget, 1962–2003. Courtesy of Baker Spring, based on numbers from the *Quadrennial Defense Review* report, 30 September 2001.

Major Component Shares of the Federal Budget, 1962–2003

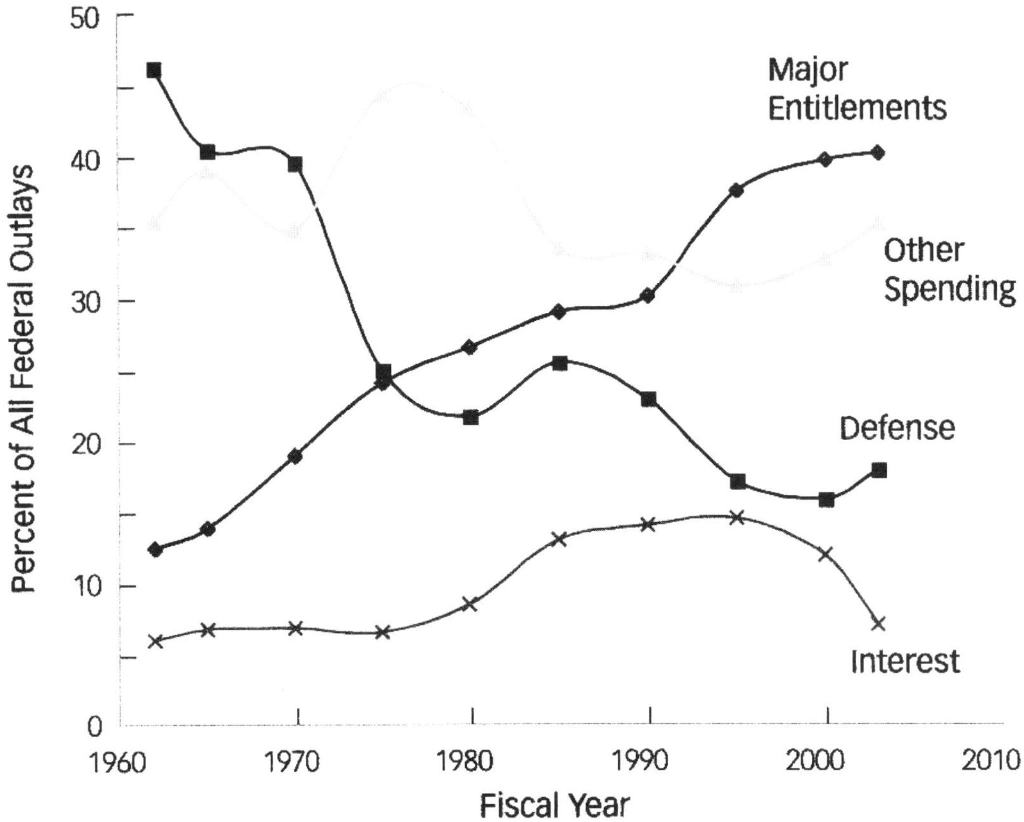

Fig. B.2. Major Component Shares of the Federal Budget, 1962–2003. Courtesy of Baker Spring, based on numbers from the *Quadrennial Defense Review* report, 30 September 2001.

GDP Shares of Major Components of the Federal Budget, 1962–2003

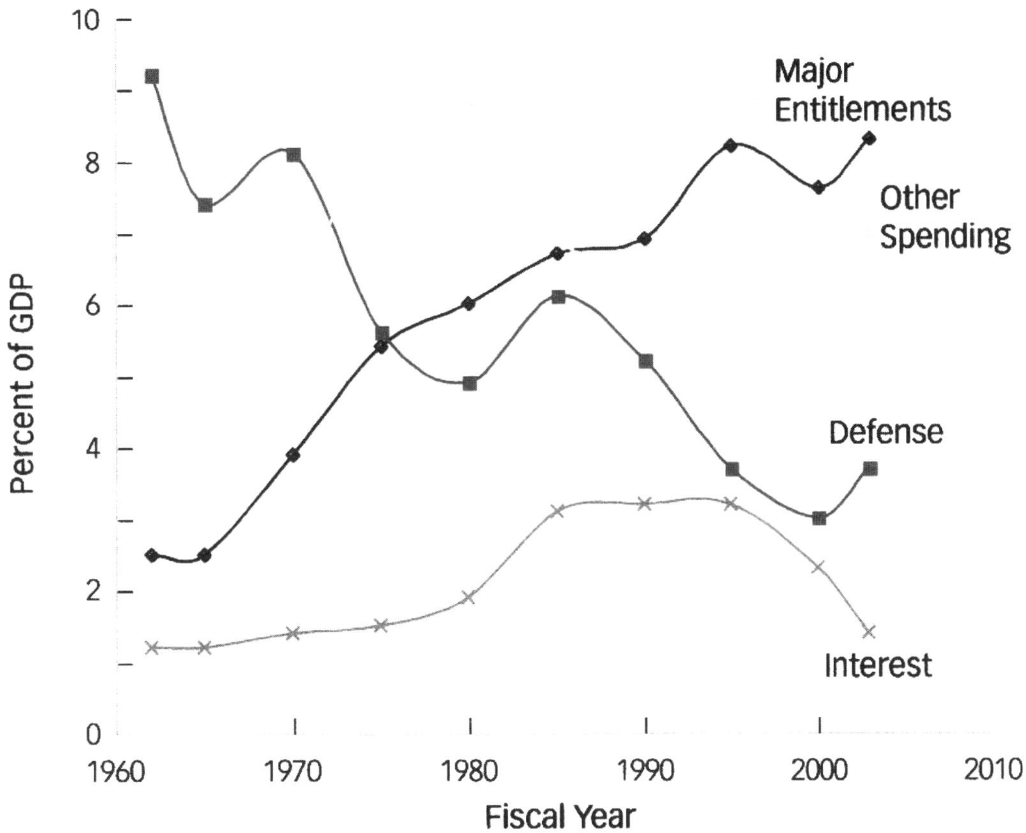

Fig. B.3. GDP Shares of Major Components of the Federal Budget, 1962–2003. Courtesy of Baker Spring, based on numbers from the *Quadrennial Defense Review* report, 30 September 2001.

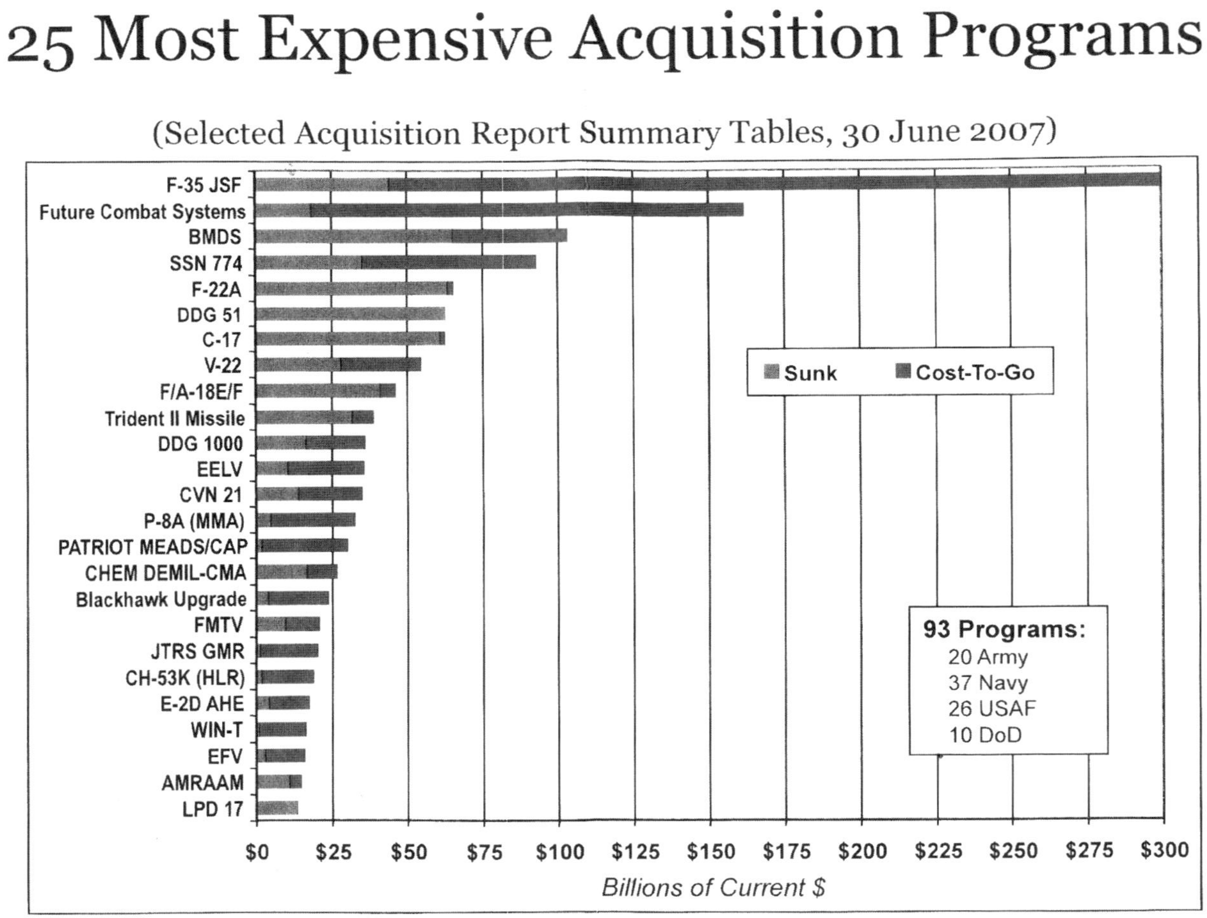

Fig. B.4. 25 Most Expensive Acquisitions Programs. Reprinted from US Fixed-Wing Modernization Plans: Near-Term Choices, presented by Steven M. Kosiak and Barry D. Watts, 19 September 2007.

Appendix C

United States Air Force History

American Airpower Strategic Thought—1901–2008 The Eternal Search for Security through Standoff and Precision

There is nothing sacred about any historical scheme of placing events into periods of time, but it is usually deemed important to make the study of history manageable. The history of airpower is no exception. This essay is largely confined to the American experience with airpower, and especially that of the US Army Air Service, Air Corps, Army Air Forces (AAF), and finally the United States Air Force (USAF). The treatment is divided up into the various ages of airpower, beginning with its origins in World War I. The next is the First Pursuit (Fighter) Age from 1919 to about 1927, followed by the Industrial Web era and the rise of the bomber, then the outbreak of World War II. Though much of the literature of that conflict is about strategic bombing, it really was an age of balanced airpower including tactical (support of the surface forces), mobility, and special air units as well as those equipped with large bombers. It also was the heyday of naval aviation.

Airpower everywhere was the child of the industrial revolution, and the first airplanes appeared just before the First World War, itself a major trauma in the story of western civilization. World War I had been followed by extreme budgetary austerity for air forces, but that was not to be the case with the second. Instead, the nuclear weapon combined with the approach to true intercontinental airpower made it seem possible to achieve security with national economy—so the air forces were well supported from 1945–1960, while the surface forces suffered short funding. The thought was growing that airpower was more for preventing war than for fighting it. The bombers were the heart of the air forces until about 1960, and then came the Second Fighter Age. The dominance of the bombers was eroded by the coming of intercontinental ballistic missiles (ICBM), Submarine-launched ballistic missiles (SLBM), and the apparent need for increased tactical aviation in Vietnam and elsewhere. However, no general war erupted, and the Cold War stalemate persisted for many years until the USSR collapsed in 1991. Following the collapse of the wall, the airpower story entered an era of uncertainty, and the age of balanced expeditionary forces.

Within these periods, the story is organized around time, money, theory and doctrine, the air mission, vision, and strategy. (In the Air Force, strategy is usually thought of as the art of relating the means to the ends or objectives.) To some degree the entire experience was conditioned by the horrors of trench warfare in World War I and the wastage of whole generations of young people. Airpower thought searched for security through standoff and precision to achieve peace and security at a less fearsome price.

The Origins: WW I, 1916–18

The Wright Brothers envisioned a military role for their invention from the beginning of flight in 1903; the Italians used airpower in combat as early as 1912. In the American case, the Navy used it at Vera Cruz, Mexico, in 1914, and the Army's First Aero Squadron did so in Mexico in 1916. However, its development in America did not keep pace with that in Europe—clearly the threat to American borders was not nearly as serious as was that among the European powers.

Budgetary Plenty—Time Shortage

Thus, by 1914 when the war started, many of the European powers had pulled far ahead of America in aviation organization and technology. Following the United States' declaration of war on Germany in 1916, the objective—at first—was that of guaranteeing the freedom of the seas. Money became plentiful; time was now the short commodity. In fact, almost all of the aircraft used by the Americans in World War I were of European design and manufacture, though the Navy did get some US-designed and -built flying boats into action.

Theory and Doctrine

Simply put, "Theory" is a body of ideas about how air forces should be organized for war and employed in war; "Doctrine" is a body of ideas about how air forces should be organized for war and employed in war with the formal approval of the highest authorities in the organization. Until the Armistice of World War I, it was premature to speak of air theory and doctrine. Insofar as it was thought of at all, it was an adaptation of cavalry roles and missions, starting with reconnaissance. In the first couple of years, the aircraft were not designed for any particular mission. The initial demand for specialization and the achievement of air superiority came from Army generals who wanted to deny enemy observation aircraft the ability to spy on them, and to make the air safe for their own aircraft so they could get information on enemy movements and for directing artillery fire. By World War I, the technology of artillery had made a quantum leap from what it had been in the American Civil War. However, despite increased range and lethality, artillery still could not engage in indirect fire—without seeing the target—as effectively as direct fire. In both the Army and Navy context, the range could reach so far beyond the horizon that the target could not be seen from ground observation posts, and this made balloons and airplanes particularly valuable to surface forces. Spotting from such high places could thereafter radically improve the accuracy of fire.

Engine power also increased rapidly. The engine in the original Wright Flyer generated about 12 horsepower (hp). In the early part of World War I, most airplanes had engines of 80–110 hp, but by its end the 12-cylinder

Liberty Engine was rated at 400 hp. This permitted the specialization of pursuit aircraft and the carriage of weapons.

Greater engine power also permitted the development of bombing aircraft, sometimes called "strategical." Strategic bombing is attacking resources not involved in current battles and is meant to affect the outcome of the campaign or war directly. In reality, these aircraft were dedicated to no more than what we would call tactical interdiction today. Interdiction promised great benefits by the delay or destruction of men and materiel moving toward the battlefield or laterally behind the lines. However, during this era, the Germans were already carrying out what came to be called strategic bombing on England—using airships at first, but later two- and four-engine bombers. The Germans also had developed a specialized ground attack aircraft before the war was over. The British established a strategic bombing unit in response, but it was just getting rolling when the war ended. Airlift and aerial delivery were used, albeit rarely. Brig Gen William "Billy" Mitchell and others observed the primitive forms of all the main missions that came to be used in World War II.

Mission

The mission of all air forces in World War I, for the most part, was support of the ground forces (later labeled tactical air), principally with reconnaissance and artillery spotting, and later with efforts at gaining air superiority. Mitchell himself came back to the United States much more concerned with gaining a separate air force and tactical aviation than with strategic bombing—the latter came a little later.

Vision

There were wartime visions of covering Germany with clouds of aircraft and also thoughts that there should be separate air forces. Pershing himself was sufficiently concerned to prohibit the use of the word "strategical" in the Army Air Services. He brought in his classmate, Mason Patrick, to bring some discipline to the Airmen whom he perceived as being good men running around in circles, and to make them understand they were auxiliaries, and the infantry remained the queen of battle—the primary force.

Strategy

Apparently, Pershing's grand strategy was to keep the US forces united and to demonstrate their decisiveness in battle. This would establish the status of the United States as a great power in the minds of European politicians. Insofar as there was an Airman's strategy, in Mitchell's case at least, the idea was to convey that airpower should be centrally commanded by an Airman at the theater level, and to some extent he demonstrated that at the 1918 Battle of St. Mihiel.

The First Pursuit Age: 1919–26

The United States generally came away from World War I persuaded that the Old World was corrupt and not inclined to pay its war debts. It was also horrified at the degree to which it had incurred a national debt. Politics moved away from the progressive prewar ideas, and the Old Guard of the Republican Party was in charge, absolutely determined to balance the budget and reduce taxes. To do that it was to follow a policy of isolationism, especially in regard to Europe.

No Time Crunch—Extreme Budgetary Austerity

The United States' industrial and financial machine was just getting rolling when the Armistice came (unexpectedly to many). Following the Armistice, there was an immediate reaction to go back to minimal Army forces, including the Air Service. Accordingly, most officers who were not discharged lost rank. The extreme austerity applied most heavily to the ground forces. Additionally, despite the Navy being seen as America's first line of defense, the funding of the naval program of 1916 was brought to a screeching halt. Soon the Washington Naval Conference was convened to limit the navies of the great powers. What followed was the scrapping of many ships, including some new hulls nearing completion. But it is fair to say that the Army was starved more than the Navy was. Also, since the time of Oliver Cromwell in the seventeenth century, at the latest, armies were often thought more antithetical to democracy and more conducive to aggression than were navies.

Time was abundant. At the Air Corps Tactical School in those days, the stress of the curriculum was minimal, and Wednesday afternoons were free for horseback riding and picnicking—the school pleaded for funding for a stable, claiming that riding improved the hand-eye coordination so essential to pilots. At war's end, Henry Arnold went from full colonel back to major; both George Kenney and Carl Spaatz, who had won a Distinguished Service Medal (DSM) in combat, were among an entitled few to retain their wartime rank. Despite being seven classes behind Arnold at West Point, the DSM enabled Spaatz to outrank Arnold, briefly, by virtue of date of rank.

Theory and Doctrine

The Air Service was subordinate to the Army general staff, and General Pershing was no fan of a separate air force. Pershing had commanded the Punitive Expedition in Mexico in 1916, led the American Expeditionary Force in Europe during the war, and was the Army chief of staff in the early 1920s. (He died in 1948 and was an important influence almost up to that point. Spaatz had tried to take him for a ride in an airplane during the Punitive Expedition but the good general refused.) Thus, to get any funding at all, Airmen first had to take care of those functions that directly served the ground forces. In fact, though, Brig Gen Billy Mitchell had returned from Europe preaching a line for a separate air force, albeit for some independent missions, he was committed first and foremost to what came to be called tactical airpower.

Mitchell, who had commanded the combat portion of US air forces in the war, returned to the United States to serve as the assistant chief of the Air Service under Maj Gen Mason Patrick. In 1919, Mitchell asserted the way to win command of the air was through winning the air battle—pursuits (fighters). Still, in speeches at the Army War College in 1921 and 1922, he did allude to the potential of the mission that came to be called strategic bombing. But in the early 1920s, his main argument had to do with the obsolescence of naval power and the ability of one air force to defend both coasts—the proposed main function.

Mitchell served as the assistant chief until his court-martial in late 1925, and soon after he resigned his commission rather than accept the punishment meted out. (Efforts were made later to reverse the verdict of the court-martial but that failed, and more recently an effort was made to promote him posthumously to the rank of major general, but that also failed.) One outcome of the court-martial was the Air Corps Act of 1926, the name Army aviation operated under until the eve of World War II when it became the AAF, and then on 18 September 1947, the USAF.

Mission

The mission of the Army, under the postwar isolationist foreign policy, was to serve as the second line of defense in case the Navy was unable to stop a hostile deployment and an invasion came overland from Canada or Mexico. The probability of invasion was so low that it could not be used to pry money out of a parsimonious Congress bent on balancing the budget and forcing the Allies to pay their war debts—Finland was the only country in Europe to pay its war debt. The Army had a coastal defense mission as well, but the Navy was not ready to concede that mission, especially in the face of Mitchell's arguments—Mason Patrick was able to get a concession out of the Army general staff that the Air Service might have some independent missions prior to the engagements of the land forces.

Vision

The vision of the country was that there would be no war in the foreseeable future, and in any case, the US involvement in World War I had been a mistake and there would never again be an overseas deployment of US forces. The vision of the Airmen was that airpower was the wave of the future. However, in the early twenties the notion in America was that command of the air was achieved by pursuit—that made the First Pursuit Group the elite outfit of the Air Service based for most of the interwar period at Selfridge Field, Michigan. As bombing and ground attack remained viable air options, the 2nd Bomb Group and the 3rd Attack Group were maintained throughout the interwar period. Ultimately, though, a third service was the ultimate vision of the Airmen, especially Mitchell, but also Mason Patrick and Benjamin Foulois,* albeit strategic bombing was not yet used as the justification.

Strategy

Insofar as the United States had a grand strategy, it envisioned staying out of foreign wars. It is probably fair to say that the United States was much more isolationist toward the Europeans than it was toward Asia. A set of contingency plans had been under development before World War I; one called War Plan Orange became probably the most important after the war. War Plan Orange envisioned a two-sided war between the United States and Japan, and essentially became the strategy of the naval forces in the Central Pacific under Fleet Adm Chester Nimitz in World War II. It was one of several contingency plans assigned a color—black for Germany, red for England, and finally rainbow for a war of coalitions.

The Airmen's strategy was to use the Navy's most famous philosopher, Adm Alfred Thayer Mahan, against the sea service. The great thinker preached that he who commanded the sea commanded the world, and the way to do that was to keep the battle fleet concentrated under all circumstances. Once the great sea battle was won, no threat to the coasts would remain. The Air Service Airmen argued throughout the twenties that the great range and mobility of airpower could be used for costal defense far out to sea, thus, releasing the Navy from guarding the main ports on each coast. Thus, the US Navy (USN) could concentrate against any oncoming enemy. The Navy was not convinced, except for a short time under Adm William Pratt, chief of naval operations (1930–1933)—and the agreement died as soon as he retired. As a result, Airmen were unable to announce any strategic bombing strategy, or even a tactical strategy, in support of a ground army, because both implied the potential for overseas deployments—and that the whole country was dead set against.

The Industrial Web and Rise of the Bomber: 1927–41

Isolationism continued to be the core of US foreign policy almost to the end of the period. Up until shortly following the election of 1932, governmental frugality was a corollary.

Time Getting Shorter—Budget Slow Increase

Major shifts began in the late 1920s. The Great Depression that began in 1929 had worldwide consequences, and was a factor in the stimulation of aggressive foreign policies in Germany, Japan, and Italy. The Japanese began their expansions in China in the early thirties, and Hitler came to power in Germany in 1933. It was becoming clear that the postwar League of Nations

*Foulois was possibly the Army's first flyer, and became the chief of the Air Corps in the early 1930s.

was impotent, and the world disarmament movement would not succeed. While the 1922 naval disarmament treaties made in Washington, DC, had been a bright start, it was becoming increasingly difficult to expand their effect or to bring about similar achievements for air forces and armies.

The Depression was hard on America, and some of the federal money intended for public relief ended up going to the military—at first, mostly, to build up the Navy to treaty limits. There was enough Army and Navy money to fund some important improvements in aviation technologies: high-octane gasoline, engine power, monoplanes, all-metal construction, retractable landing gear, internal wing bracing, and drag reduction. Pursuit aircraft in World War I, in the best cases, could get up to perhaps 120 miles per hour (mph). By the onset of World War II, they could exceed 300 mph, and by its end, a few could speed at more than 400 mph. The most powerful engines in WW I were 400 hp. By the late 1940s, 3,500 hp engines were coming off the line for the C-97 and the B-36. Two thousand hp engines were then common in smaller airplanes.

Theory and Doctrine

From the mid-1920s onward, the Navy also began developing aviation aboard the experimental aircraft carriers *Langley*, and the much larger *Lexington* and *Saratoga*. There were growing thoughts the carriers would replace the battleships as the primary weapons system, and that organization by type would be replaced by carrier task forces. In the Army Air Corps (from 1926), there was a transition from the age of the fighters, where the new elite units became the bomber outfits. One result was the development of the B-17 four-engine bomber—named the Flying Fortress—precisely to put a defensive spin on the project. The declared theory and doctrine behind the creation of long-range, four-engine bombers was the need for coastal defense in accordance with the MacArthur–Pratt agreement in force in the early thirties. Despite the shifting tides, Airmen could not openly espouse theories of strategic industrial bombing, as that would contradict the national isolationist foreign policy—and further alienate the Army general staff that controlled the Air Corps budget. It should be noted that the Air Corps was much better funded than the other branches of the Army; notwithstanding the Airmen's laments, and neither Mexico nor Canada presented much of a threat to security.

Meanwhile, the theory for strategic bombing was under development by the instructors in the Air Corps Tactical School's "Air Force" course. There were also other courses devoted to pursuit and attack aviation, but clearly the "Air Force" presentation was the primary offering. With no hint at all of the possibility of radar, most Airmen looked upon the practicality of an effective air defense with disdain. Escort fighters were used in World War I, as were disposable "drop tanks," but as escorts would have to carry large amounts of fuel for the return home with the long-range bombers, it seemed inevitable that they would always be much heavier—and less agile—than the defending interceptors. Thus, their development seemed futile, and even the fighter advocate Claire Chennault agreed with his bomber colleagues. Perhaps through wishful thinking, the hope was that the bombers would achieve force security by flying at very high altitudes and in tight, controlled formation. By the time the interceptors got up that high and found their prey, they would be so low on fuel that they would have to go back to their nests almost immediately. Ultimately, radar changed all that.

The general staff was stingy on funding for four-engine bombers, but it did flex enough to further recognize the possibility of independent air operations prior to the contact of the ground forces and permitted the organization of the General Headquarters (GHQ) Air Force in 1935—said to be the precursor of an independent air force. The GHQ Air Force was organized into three wings: the First Wing at March; the Second Wing at Langley Field, containing all of the four-engine heavy bombers, and the Third Wing at Barksdale Field—heavy on attack aircraft and ground support capabilities. All were generally composite wings, containing pursuits, attack aircraft, and bombers. In general, the bomber elite were gathered under Maj Gen Frank Andrews as his staff at GHQ, and in the Second Wing flying the early B-17s.

For a time, the GHQ Air Force was responsible for unit training and operations, and the Air Corps was responsible for individual training and equipping the air forces. General Andrews commanded the GHQ AF, and seemed headed for higher things. He was a West Pointer and had gone on to be the theater commander in the Caribbean and then on to England after the war broke out. However, before greater things could happen, he was killed in 1943 in a B-24 accident in Iceland. The Air Corps was headed by Maj Gen Henry Arnold, also a West Pointer. When the two organizations merged into the US Army Air Forces (USAAF) just before World War II, Arnold became its commanding general and remained in charge until Carl Spaatz took over after the war. (Arnold, the only Airman in history to achieve five-star rank, died soon after retirement. Admirals William Halsey and Ernest King achieved five stars and both had wings, but neither had ever served at the squadron level in aviation.)

Mission

The official primary mission remained the support of the ground forces through air superiority and ground attack. The coastal defense mission under the MacArthur–Pratt Agreement ended with the retirement of Admiral Pratt in June 1933. Thus, the coastal defense mission was again open for grabs—though the Army still had the ground-based portion. There was an official recognition of an independent mission for the Air Corps, but that was thought to be attacking the bases and lines of communications of land invaders prior to their contact with US ground forces.

Vision

Mitchell had come back from Europe in 1919 with the vision of an independent air force, coequal with the Army and Navy, with no more than a glimmering of the notion of a strategic bombing force. That vision was probably shared, in the main, by most of the Air Service/Air Corps heavyweights throughout the interwar period. As time passed, Airmen increasingly had the vision that airpower would replace naval power as the first line of American defense and that perhaps it could be decisive in **some** air-alone campaigns. Additionally, they tended to share a conceit that the air arm was the only forward-looking service and the technological leader.

Strategy

The bureaucratic strategy of the Airmen was complicated by the fact that they had to cautiously deal with the Army General Staff, as it was aware of the Airmen's vision for a separate air force—and most staff officers were dead set against it. Thus, airpower theory and technological development had to be done cautiously often in an academic setting at the Air Corps Tactical School—and even then there were always officers from the other services attending who would report back to their own superiors.

Insofar as there was a strategy, it was focused on the development of the technology and the tactics for long-range bombing missions. The Norden bombsight was developed—it was thought—to enable accurate bombing in daylight from standoff high altitude. Long-range missions were often sought for public relations reasons, which also helped with the evolution of air navigation. Increasingly, Airmen realized that war was coming and probably the United States would not be able to sustain its isolationist policy. Thus, they hoped for the opportunity to demonstrate the decisiveness of airpower with mass, daylight, and accurate attacks on key centers of gravity in enemy industrial and military systems.

WW II and Balanced Airpower: 1941–45

After 1938, the United States began to abandon its isolationist foreign policies, though in stages—even before Pearl Harbor, it was effectively engaged in a naval war in the North Atlantic. Unlike World War I, this time there was a rather clear forewarning that the United States would be engaged in the war, and thus, the nation had a couple of years in which to get its industry and armed forces rolling. The goal became to stamp out Fascism, Nazism, and militaristic aggression.

Out of Time—Budget Plenty

It can be said that airpower truly came of age during the Second World War, as technology, theory, and tactics met in a veritable perfect storm. Armed with vast amounts of money and increased industrial capacity, the American air forces tested their theories, strategies, and equipment in the crucible of World War II. Despite the devastating bombing attacks carried out against the Axis economy in Europe, and the destruction wrought on Japan and its people in the Pacific, it was unclear to many just how effective strategic bombing had really been. Whereas, many Airmen saw the tactical campaigns of Ninth and

Twelfth Air Forces in Europe, and Fifth, Seventh, and Thirteenth Air Forces in the Pacific as the most successful applications of airpower during the war, Hiroshima and Nagasaki obscured much of that—as did the vicious postwar arguments over unification of the services and a separate Air Force. At one time during the debates, Gen Carl Spaatz of the AAF asserted that Airmen preferred one unified service—Army, Navy, Air Force under one service—but if that were not forthcoming, then two would not do. Rather, there would have to be three, to include the separate Air Force.

Theory and Doctrine

The legacy of World War I and the ideas of Giulio Douhet, Billy Mitchell, Alexander de Seversky, and the Airmen of the Air Corps Tactical School served as theoretical beacons for the AAF during this time. The successful German blitzkriegs that kicked off the European conflict were indeed noted by Airmen and all Americans, but it was clear to most flyers that the best way to end the war against the Axis powers was to strike their hearts and destroy their physical and mental war-making capacity. Giulio Douhet, an Italian army officer, was once jailed for his radical thinking. He became an icon for those people devoted to the strategic bombing idea because of his book, *The Command of the Air*. Often thought to be the apostle of population attack, he was doubtless more radical than most American devotees to strategic bombing. Alexander de Seversky, a Russian who immigrated to America, was the founder of the aircraft manufacturing company later known as Republic Aviation and reached fame during World War II with his book *Victory through Airpower*. The book was converted to a Walt Disney animated movie, usually shown at Maxwell AFB's Squadron Officer School (SOS) in the 1950s. Mr. de Seversky, himself a pilot, was about as enthusiastic supporter of strategic bombing as existed in the United States and continued to lecture to SOS students as late as 1959.

In any event, as no invasion of France could be contemplated before 1944, the only way to keep direct pressure on Germany in the meantime was through strategic bombing. It was imperative to keep the Russians in the war, and while building up for the invasion, strategic bombing (it was hoped) would serve as enough reassurance to Stalin to keep him from making a separate peace—albeit the same reasoning was used for a rationale to launch the African Campaign in November 1942.

Mission

Simply put, the mission of the AAF was to safeguard the United States and its allies by taking offensive action against the Axis powers and supporting ground offensives in case offensive air action was not totally successful. Many Airmen at the time firmly believed that Kenneth Walker's*

*Brig Gen Kenneth Walker, killed in a B-17 raid against Rabaul in January 1943, and earlier one of the instructors in the Air Force course at the Tactical School.

well-planned, well-flown strategic bomber missions would not be stopped, and the death and destruction they wrought would bring the enemy to its knees without the slaughter seen in the Great War's trenches.

Vision

The AAF vision was to have a modern air force capable of attacking the enemy heartland and gaining control of the skies. Unlike the Luftwaffe, though, both the Royal Air Force and the USAAF developed "general" airpower including strategic, tactical, mobility, and special operations air units. Additionally, the push for an independent American air force was still the dream of many US Airmen—even though the AAF was in many respects a separate branch of the armed forces during World War II. Despite the fact that some American Airmen pushed for one unified force, rather than the addition of another service, the single-service idea lacked decisive political support.

Strategy

Different successive Air War Plans Division (AWPD) strategies for the conduct of the air war against the Axis were formulated during this time, each adjusted according to the context and progress of the Allies. However, all called for the destruction of German and Japanese war-making capacity, followed by support to Allied surface forces. Germany was prioritized over Japan, and the priority of targets for affecting Axis war-making capacity was often shifted. For example, August 1941's AWPD-1, *Munitions Requirements of the Army Air Force*, called for the disruption of the German electric power and transportation system, destruction of German oil and petroleum, and undermining German morale by attacking the civilian population. By September 1942, in AWPD-42, *Requirements for Air Ascendancy, 1942*, defeating the Luftwaffe and destroying sources of German submarines moved to the top of the list. As for tactical airpower, the foundations were laid during the African Campaign, and were articulated in the War Department Field Manual 100-20, *Command and Employment of Air Powwer* (July 1943)—airpower to be centrally controlled at the theater level by an Airman; airpower coequal to land power; air superiority as the first priority, then interdiction, then CAS; and finally reconnaissance and airlift. In a ground emergency, the CAS can take priority. In general these remain the principles of tactical air doctrine.

The Heyday of the Nukes and BUFFs:* 1945–60 Containment

The period after 1945 would differ markedly from the interwar years. Whereas after World War I the United States adopted isolationism, containment of the communist threat was to be the essence of US foreign policy after the Second War. The American economy was robust, and a

*"BUFF"–Air Force jargon for B-52s, (Big Ugly Fat Fellow)

communist-predicted postwar depression never occurred. In general, after Korea, the federal budget was successfully balanced several times in the 1950s.

Time Crunch Moderate, AF Budget Plenty—Army Starved

The newest of the armed services, the Air Force, found itself with ample funds for jets, but also for tankers, new electronics, missiles, and nuclear weaponry. The Soviets tested American will directly in Berlin, and indirectly in Korea, and heightened tension by exploding its own atomic device in 1949. American political and Air Force leaders grappled with this new nuclear age, reconciling old airpower lessons to new weaponry, and relying on its nuclear hegemony to prevent wars—or failing that, to assure survival.

Theory and Doctrine

Air refueling had been used experimentally before World War II, and now was developed and expanded greatly to enable the use of jet bombers that would depend on their speed for security during penetration of enemy air space and egress. The combination of jet penetration and nuclear weaponry led many to believe that Douhet had finally been validated and that a few heartland strikes would stop wars.

At the onset of 1944, any second lieutenant in Eighth Air Force had been deemed to have a 22 percent chance of completing a 25-mission tour in bombers. More than half were predicted to be killed or made prisoners. Were a nuclear attack to suffer that kind of attrition over an extended period of time, it would certainly have exceeded the capability of the United States. Thus, the theory was with air refueling, jets, and nuclear weapons, the campaign would just require two or three missions from each lieutenant, and thus, survival chances would be reasonable, even by World War II standards.

Deterrence theory remained increasingly central to the strategies of the United States and its Air Force, arguing that the risk of nuclear annihilation would dissuade enemy aggression. In this way, many air leaders saw airpower—in both active and passive forms—as maintaining a world balance of power and thereby sustaining international peace. The Korean War demonstrated that nuclear weapons alone would not be enough to sustain the emergent Containment Policy. In light of the Korean War, others began to see the nuclear stalemate of the Cold War as not empowering peace but as at least limiting wars below the nuclear level and preventing direct conflict between the superpowers. The Air Force doctrinal documents of the 1950s heavily reflected the World War II experience—strategic attack is primary, air superiority is essential, tactical airpower should be centrally controlled at the theater level by an Airman, and interdiction is preferred over CAS, except in an emergency. As the decade wore on, ICBMs, and later SLBMs, gained a role in nuclear deterrence.

Mission

The Cold War was just beginning. Despite its marked differences from the recent war, the Air Force, in some ways, seemed to be a service at war. Thus, the mission of the Air Force reflected that of the AAF during World War II. The new USAF was charged with being able to attack the Soviet Union in order to destroy its war-making potential, and to defend the homeland from enemy air attacks. It was also responsible for supporting the other services with airlift, reconnaissance, and tactical support.

Vision

Before retiring, Hap Arnold's vision for the Air Force was that of an "organization of people, ideas and machines with the purpose of delivering devastating blows to the enemy, empowered by the technology of a nation excited about aviation." Later, Curtis LeMay described his vision of the new Air Force similarly, stating that the "fundamental goal of the Air Force should be the creation of a strategic atomic striking force capable of attacking any target in Eurasia from bases in the United States and returning to the points of take-off." For all the prominence of SAC, tactical airpower in its various forms was not ignored in the Eisenhower years. The Tactical Air Command (TAC) gained a nuclear capability in those years, and many new systems and weapons were brought into its inventory. In fact, TAC deployed its Composite Air Strike Force to Turkey in response to the Lebanon Crisis of 1958, among other operations.

Strategy

As the United States ended one war and entered the Cold War, the country faced a constrained economic situation. Pres. Dwight Eisenhower understood that the large military, with its giant-sized budget, could not be sustained, so he and his advisors looked for ways to ensure US national security and relative international peace without bankrupting the nation. One strategy was a continued emphasis on collective security, seen most notably in the sustenance of NATO, which had been created in 1949 under the Truman administration.

Another strategy was termed *New Look*, in which the United States would rely on relatively inexpensive nuclear weaponry and airpower, rather than a large army or navy, to protect the nation. From this flowed the Massive Retaliation Strategy, in which the capability of wiping out the Soviet Union would deter its leaders from directing an attack on the United States. Secretary of State John Foster Dulles seemed to threaten such a response even in cases of aggression in peripheral areas. Some Airmen found such a strategy as imprudent, preferring instead a Counterforce Strategy, in which the Air Force would concentrate its attacks on Soviet forces and nuclear stockpiles; and later, the Flexible Response Strategy, in which a massive nuclear exchange was not to be automatic in the case of aggression. By the end of the period, Finite Deterrence Strategy, better known as

Mutually Assured Destruction (MAD), and ultimately known as a Second Strike Strategy, was advanced by the USN. Finite Deterrence and the Second Strike Strategy argued that the superpowers should maintain just enough nuclear weaponry to guarantee unacceptable losses by each should war break out, thus ensuring neither would ever take the first step. This was thought to be stabilizing, whereas weapons useful for the First Strike would be destabilizing.

In general, up until the mid-1950s, the Navy was an unwavering foe of strategic attack. However, as it dawned on many sailors that nuclear powered submarines would be absolutely secure and possibly armed with solid propellant rocket-powered missiles tipped with miniaturized nuclear weapons, they saw that the Navy could acquire an important part of the nuclear pie with a new mission. This was all the more potent since the advance of ICBMs was making the bomber force ever more vulnerable at its airfields. For the bombers, airborne alert was inordinately expensive, and only a small fraction of the whole alert force could be kept airborne at any one time. Thus the marriage of nuclear weapons and nuclear powered submarines came along at a particularly opportune time. Such a force was said to be particularly valuable, because its accuracy could never be good enough to take out enemy missile silos, so it could only be used as a Second Strike against cities.

The Coming of ICBMs, Parity and the Second Fighter Era: Counterinsurgency, False Dawn of Détente, 1960–91

Eisenhower recognized that the nuclear hegemony of the United States could not last forever. Thus, stirrings of the idea of a more flexible defense policy were afoot among Army officers and even some Air Force generals such as Lauris Norstad—until recently, Norstad was the only Air Force officer to command United States European Command and NATO. The Korean War had been demonstration enough that nukes would not guarantee eternal peace, and that combat was likely to occur below the nuclear level for many years to come. Even during the heyday of the so-called "Bomber Barons" of the 1950s, the F-100, M-61 aircraft Gatling Gun, C-130, C-124, and a string of conventional bombs *were* developed. None, save the C-124, had much of anything to do with nuclear warfare or deterrence. The AIM-9 Sidewinder air-to-air missile got its first kills over the Straits of Taiwan in 1958.

Actually the concentration on nuclear operations of the 1950s can be overstated. The Air Force had more fighters on the line in 1955 than in 1965, though many were not in TAC. Additionally, the Navy was authorized the four ships of the *Forrestal* class in the 1950s, all about the same size as the US super carrier that had been cancelled under the Truman Administration.

The mythology surrounding the 1950s makes the Bomber Barons appear more powerful than they were. They were supposedly led by Carl Spaatz at first—but he also had fighter experience with three air-to-air kills in World War I combat and had commanded the 1st Pursuit Group when it was the elite unit in the Air Service. Spaatz was succeeded as chief of staff by Hoyt Vandenberg who cut his spurs as a ground attack instructor pilot in the 3rd Attack Group, and led one of the greatest tactical air campaigns in history as head of the Ninth Air Force as it moved across Europe. Vandenberg was succeeded as chief by Nathanial Twining, who also started out as an attack pilot in the Third Wing, not getting into a bomber unit until he was a senior officer. Twining was succeeded by Thomas White, who was chief at the end of the Eisenhower administration, had spent most of his junior years in the attaché business, and also did not get into bombers until he was very senior. All four of those chiefs were West Pointers. By January 1961, a new administration came to power determined to build a new "Flexible Response" defense policy.*

Time Crunch Moderate—Budget Apparent Plenty

Soon after taking office, the new president, John Fitzgerald Kennedy, suffered a considerable humiliation with the Bay of Pigs fiasco. That was followed the next year with the Cuban Missile Crisis, and it seemed that the Cold War had reached a climax. It is now known that both sides made concessions, and what seemed like high chances of combat between the superpowers disappeared. Soon after, the Kennedy administration became increasingly involved in Southeast Asia. Painting the situation was the belief that there was a challenge that had to be met under the Containment Policy. That policy, in force since the late 1940s, desired to contain further communist advances in the hope that time would moderate the aggressiveness of its revolutionary strains and at the same time avoid another world war. The extension of the Containment Policy to Southeast Asia in the 1960s assumed that the Vietnamese communists were merely the tools of Moscow, and therefore, it was necessary to arrest their advance southwards. Ultimately, the result was the Vietnam War and a challenge at both the tactical and counterinsurgency levels.

The election of 1960 had been one of the closest in American history, and the economy was still in fairly good shape coming out of the prosperous and relatively peaceful Eisenhower years. Cultural conflict loomed as a new generation of youth came on the scene with different standards and expectations. The draft continued to help with the manning of the still massive armed forces. The huge expenses associated with maintaining a large military competed with the social needs in the homeland. The perceived humiliation of the USSR in the Cuban Missile Crisis was said to have induced a "never again" attitude among the decision makers in Moscow and stimulated new defense spending in both America and Russia. The Soviets had launched Sputnik in 1957—deemed an extreme humiliation

*Some F-100s did wind up standing alert with tactical nuclear weapons, but they were basically a day air superiority fighter designed to succeed the F-86 of Korean War days.

for America. With an urgency to take some action, and without much forethought, Kennedy committed to putting a man on the moon within a decade.

Theory and Doctrine

Clearly, the nuclear dominance enjoyed in the Eisenhower years had waned and the deterrence theory had to be modified. As noted, what finally emerged was a Second Strike Theory that permitted a leveling out of strategic nuclear forces and the diversion of funds to other foreign and domestic policy needs. The notion was that massive nuclear superiority was not needed to deter. Rather, as long as each side could guarantee the survival of enough of its nuclear striking power after an attack from the other side to produce unacceptable damage to the initiator in a second strike, then the two sides would be mutually deterred at the strategic level—regardless of what was happening in conventional and insurgency battles around the world. That also meant that the various contests would be decided without nuclear weapons, and thus the conventional capabilities had to be restored to handle such things. In the Air Force, that led to what was seen as a Second Fighter Age and the rise of the fighter generals.

When Gary Powers was shot down in a U-2 over the USSR in the last year of the Eisenhower Administration, further emphasis on the development of space for reconnaissance purposes followed. Though the U-2 flights had been under way for some time, it was clear they were a violation of international law. However, the Soviets had not asked for overflight permission for their Sputnik in 1957, and that enabled the United States to initiate a Freedom of Space doctrine. By the time Powers was shot down, the United States was on the verge of accomplishing the same function from space. The new intelligence permitted better security and verification of arms control agreements, and also was conducive to some leveling out of spending on strategic attack weapons.

In this timeframe, the Army was able to undertake massive development of air mobility capabilities based on helicopters. The fortunes of special operations forces, both in the Army and in the Air Force, took a sudden turn for the better. The Army special operations units at Fort Bragg, North Carolina, and the Air Force at Hurlburt Field, Florida, were rapidly developed for operations beneath the level of major conventional combat operations, and were often covert in nature.

Vietnam led to further weakening of any tendency to rely on nuclear power for anything but the passive role of deterrence—and led to the end of the eight-year spell of Democratic presidential rule in Washington. Republican Richard Nixon won the election of 1968, and again in 1972. He and Secretary of State Henry Kissinger visited China and simultaneously brought about a rapprochement with the USSR so that a departure from Vietnam was deemed possible. It seemed to witness a dawn of an era of détente—in which the Cold War would be moderated if not ended.

It was a false dawn. Soviet incursions in Afghanistan and adventures around the Horn of Africa, combined with America's apparent paralysis induced by the Watergate Affair, caused the continued reliance on the second strike on both sides.

Many devotees to strategic bombing credited the settlement of the Vietnamese Peace Talks to Linebacker II—an 11-day attack on North Vietnam during the Christmas season of 1972—and the final vindication of the B-52s and strategic bombing. Others thought it said nothing about strategic bombing, because throughout, practically all of the classical strategic targets were in China and the USSR, and could not be attacked. In reality, it was much more complicated; with many Americans believing the nation could not survive many more victories like the one in Vietnam—as the North Vietnamese achieved practically all of their goals by 1975.

Meanwhile, the Soviet competition with the United States and the seemingly endless campaign in Afghanistan, seriously weakened the USSR, to the point where it collapsed without a nuclear war, or a direct conventional campaign with the United States and NATO. Quite unexpectedly to US strategic planners, the Cold War was over and the heyday of SAC had passed. The main reason for the creation of a separate Air Force seemed gone.

Mission

DOD leaders declared the mission of the US' nuclear forces was to deter a nuclear attack on the United States or its allies, as well as deter a major conventional attack against either. Further, the Air Force was to support the other services with airlift and tactical air support. These missions held true during the Cold War, but as the Soviet Union began to crumble, some in the Air Force felt the service was having an identity crisis. Deprived of its longstanding enemy and stripped of its leading wartime role by the Goldwater–Nichols Act—regional or combat commander's fight; services organize, train, and equip—a group of Airmen in the Pentagon wrote a white paper, titled "A View of the Air Force Today," contending that the USAF was obsessed with technology and had lost its purpose. About the same time, Carl Builder argued a similar case in *The Icarus Syndrome*.

Vision

Donald Rumsfeld, during his first turn as Secretary of Defense (1970s), said the United States needed "a well-protected, second strike force to deter attacks on our cities and people . . . a capability for more . . . measured responses, to deter less than all-out attacks . . . [and] essential equivalence with the USSR." Partly in response to concerns expressed in "A View of the Air Force Today," Air Force Secretary Donald Rice published in 1990 the first iteration of Global Reach–Global Power, describing the role of the Air Force in national security and the increased importance it would play in the future. Finally, Chief of Staff Gen Merrill McPeak announced the first official USAF vision in 1991: "Air Force people building the world's most respected air and space force . . . global power and reach for America."

Strategy

Secretary McNamara's Flexible Response strategy had called for specialized weapons, hardened weapons deployment, and secure command and control that would permit measured engagements against hostile military forces rather than all-out spasm attacks against cities. Additionally, it called for the capability to attack in waves, whether the targets were cities or enemy forces, and aimed at a nuclear stalemate rather than strategic superiority as many air leaders advocated.

In the Vietnam War, the Air Force strategy began with the gradual increase of pressure on the North Vietnamese in order to defuse any potential conflict with the Soviets or Chinese. It ended with the aggressive bombing of North Vietnamese vital centers in conjunction with peace negotiations.

Later, as part the 1991 Gulf War, air strategy planning occurred on the Air Staff in Col John Warden's Pentagon office. The planners working for the US Central Command joint force air combat commander completed it. Though the strategy was said to go for the center of Warden's "5 Rings"—attempt to cut off the head of the snake—it almost simultaneously provided for achieving air superiority over Kuwait, the preparation of the battlefield for the upcoming invasion, and ultimately close support for the ground penetration itself. Afterwards, this was called "parallel attack," in contrast to the gradualism of the long Rolling Thunder's bombing effort (1965–68) against North Vietnam. Though thousands of fatalities had been predicted, the four-day ground campaign that followed resulted in only 148 deaths—14 of whom were aboard a single AC-130 that was shot down with an infrared missile. Despite that loss, though, it was a fine hour for American Airmen.

When Warden was a student at the National War College, he wrote a thesis later published as *The Air Campaign*. In it he posited a model of an enemy arranged in five rings, with leadership at the center (the head of the snake) and with the fielded forces out on the perimeter. He suggested that victory might be had more easily and cheaply by directly going at the center bypassing all the outer rings. The book became exceedingly popular in Air Force circles, especially by advocates of strategic bombing.

The Era of Balance, Precision, C3, and Space: 1991–2001

The First Gulf War occurred almost simultaneously with the fall of the USSR and the Warsaw Pact—the United States little anticipated the latter events. It was a huge, and for a change, a pleasant surprise. Suddenly the longed-for end of the Cold War was upon us. The peace dividend that had been dreamed of off and on since Hiroshima seemed finally at hand. The Air Force and the other services were at Cold War strengths still when the Gulf War erupted, but all involved knew that a massive drawdown was imminent. The budget had not been balanced since the Eisenhower administration ended 30 years before. Many domestic needs had long been unfunded or underfunded. Only one superpower remained; peace, prosperity, and globalization seemed certain for a long time to come.

Time Abundant—Budget Peace Dividend

For the Air Force, all that implied major reorganization as well. The fall of the Soviet Union and the emergence of a weakened Russia left the United States as the sole global superpower. The resulting international pressures demanded a review of Air Force organizational structure in light of the post-Cold War era.

Theory and Doctrine

The bread and butter function of the Air Force, strategic bombing, had long declined, though it had not been abandoned as it was thought essential for its passive utility of deterrence. Even the urgency of deterrence seemed much diminished in view of the reduction of the Soviet threat. Air superiority remained primary in doctrine, and interdiction and ground support were both essential, though arguments with the Army arising from its AirLand Doctrine and its domination of the placement of the Fire Support Coordination Line (FSCL) continued. In general, the FSCL is associated with linear warfare, and it is set by the ground commander at some distance in front of the troops—implying that all fires inside that line had to be coordinated with the ground commander to lessen the chances of fratricide on all sides. Air efforts inside the FSCL were greatly slowed by the coordination requirement, making it difficult to act in time to kill fleeting targets.

Global mobility through airlift and air refueling remained as essential as ever, and long-range strike with conventional weapons was still thought required, though not necessarily associated with industrial targeting—or with big bombers. Important modifications came out of Desert Storm in the concepts of parallel (instead of sequential) attack and effects-based targeting. Instead of mere target destruction, the Air Force moved to an approach that sought only effects—like shutting off electrical power or stopping operations at an airfield. It was believed that shorting electrical transmission lines would be quicker and easier, compared to the destruction of dams and generators, and the effect would be easier to repair later; also, cratering a runway would be quicker and relatively easier instead of destroying the entire airfield.

Also, since future enemies were uncertain, the service changed from a threat to a capability-based theory. A major implication of the end of the Cold War was that the Air Force would no longer be forward–based, but rather become an expeditionary force. The combination of air mobility, air refueling, and long-range strike became vital in speeding up the application of force, because crisis after crisis has shown that the earlier force can be applied, the less total force will be required.

Mission

In the post–Cold War era, the Air Force's mission became: (1) to prepare and equip air, space, and cyber forces for the utilization of the combatant commanders to achieve air superiority, interdiction, CAS, tactical airlift, reconnaissance, and special operations; (2) to provide long-range missile and bomber forces for employment by USSTRATCOM and the other combatant commanders; and (3) to prepare and equip air defense, long-range airlift, and air refueling forces.

Vision

The slogan "Global Reach–Global Power," as a basic vision denoted that Airmen were best equipped to deliver rapid and decisive effects, both lethal and nonlethal, from the CONUS to anywhere on the globe—and to do so with a much smaller Air Force. As the capabilities of space became better understood, "Global Vigilance" was added to the slogan.

Strategy

The Air Force adopted a strategy to reorganize in two ways: (1) to consolidate functions of various commands to bring about savings; and (2) to change organization and training to enable the transition from a forward–based to an expeditionary force. Ultimately, this concept resulted in combining the logistical functions of the former Air Force Systems Command (research and development) and Air Force Logistics Command (supply and maintenance) into a single Air Force Materiel Command. It further combined the lethal airpower of the old TAC and SAC commands under ACC, based at Langley AFB. Below that level, it eliminated the remaining air divisions and strengthened the lesser units, as required.

As the threat was no longer easily predicted, forward basing was largely abandoned and units were brought back to the homeland with the idea that they could be redeployed in any direction, as required. The surviving units were divided into 10 AEFs, two of which were deployable as a matter of routine, and the others under wartime requirements. Most of the mobility assets were not a part of the 10 AEFs, due to their limited numbers, and they had to support all of the AEFs, as well as the other services. As a part of the process, bases were to be closed and total force was to decline to about one third the size it had been at the height of the Vietnam War. The drawdown was interrupted for a time by the Gulf War in 1990–91, but resumed shortly thereafter.

During the remainder of the 1990s, airpower seemed to be the nation's weapon of choice. It was used with local ground forces in Bosnia, practically alone in the air war over Serbia, and again with local ground forces in Afghanistan after the 2001 attack on the American homeland. Though airpower was not formally used in preparation for the 2003 Iraq ground attack, it had nonetheless effectively prepared for that campaign for a decade in Operations Northern and Southern Watch—air superiority was already firmly established over much of Iraq when the three-week march on Baghdad began. Through it all, the reach of air forces had been gradually extended by air refueling even for small airplanes, and each sortie was made more effective by the increasing use of precision weapons, thus reducing the total number of weapons required by an order of magnitude.

Looking back for a moment, the increase in strike efficiency through precision guidance has a longer history than is often assumed. Legend has it that the first precision guided munitions (PGM) were developed by Genghis Khan. In the process of besieging a Chinese city, he suggested to the town fathers that if they would pay him a tribute of 1,000 pigeons and 1,000 cats, he would go away peacefully. The town dignitaries could not believe their good luck and quickly gathered up the tribute and delivered it outside the walls. Instead of going away peacefully, the good Genghis Kahn had a scrap of cotton treated with incendiary material attached to them all, and set it alight. Then the cats and pigeons were released to return to their cellars and attics in the town and promptly burned the place down.

More provable is the American case. The Azon and Bat guided weapons were employed with some effect late in World War II, using radio guidance in the former and radar in the latter. Azon and a better version Razon were used by the Air Force in the Korean War with some effect, but low reliability. The Navy developed the radio-guided Bullpup that was used early in the Vietnam War, and both laser guidance and electro-optical guidance came on towards its end—with excellent effects repelling the Easter Offensive of 1972. Around 10 percent of the weapons dropped in Desert Storm were PGMs. The GPS/inertial guidance was applied in bombs during the 1990s, creating an all-weather weapon, and one that could reliably guide itself to the target coordinates without the assistance of the aircrew after release. Up until then, infrared guidance was required for weapons to be used at night, and even that suffered from obscurants in the atmosphere. Since the 1990s, guided weapons have been increasingly used, not only for their economy and effectiveness but also because the precision has greatly reduced collateral damage.

When increased standoff capabilities were married with precision, enabled by a worldwide electronic command and control system, even the reduced force was more powerful than could have been dreamed of by the Airmen of yesteryear.

The Age of Strategic Extension: 2008–23

As things stood in 2008, the world ahead seemed as uncertain as it ever had been, notwithstanding the information revolution. The Second Gulf War morphed into a counterinsurgency campaign that put a seemingly endless strain on the national budget. The price of energy moved radically upward, having negative effects all around the world. The road to further globalization and the associated efficiencies seemed blocked by various impediments. Deadly effects seemed available to nonstate actors of every stripe. American domestic needs were going unfunded again after the balanced budgets of the 1990s—the national debt was rising to unprecedented heights. Rising productivity and incomes in China, India, and even Russia, were putting additional stresses on world resources. Developments even seemed to be appearing in nature itself. All this poses a problem for American strategists who now must build their plans on a whole new set of assumptions about the path that lies ahead.

Appendix D

Quick Look (Background) Papers

CADRE Quick-Look

Catalyst for Air & Space Power Research Dialogue

APRIL 2006

Islands in the Stream: The Advent of the One-of-a-Kind Air Force Base

John L. Conway III

Issue. During the Cold War each Air Force base had a Survival, Recovery, and Reconstitution (SRR) Plan, detailing dispersal sites and the procedures to reconstitute aircraft and personnel after a nuclear laydown. To the planners, the end of the Cold War meant the end of the threat of direct attack on the American homeland—SRR plans gathered dust.

Following 9/11, new fears of a direct attack on the homeland drove new threat assessments, created Threat Working Groups focused on terrorism, and renewed the Air Force's emphasis on Chemical, Radiological, Biological, Nuclear, and Explosive (CRBNE) preparedness. Once again, we began to think in terms of direct attack on America.

Conducting its own devastating attack on the American homeland, Hurricane Katrina struck a serious blow to Keesler AFB, Mississippi in August 2005. Training missions were disrupted and trainees were dispersed to other bases; damage to base housing topped $300M; and, in an ironic twist, the base's famed WC-130 "Hurricane Hunters" became the hunted and were forced to flee.

While a hurricane evacuation plan is a fixture of any East Coast or Gulf Coast Air Force Base, it involves only dispersal and a return to home station. But what happens if a base is seriously damaged and its primary mission is no longer viable there? Where is an alternative work site? Can other bases pick up the mission until the base is functional again? What about relocation of its trained workforce?

Unlike the plethora of detailed terrorist threat scenarios, there are no in-depth assessments of an air base's susceptibility to natural disasters. We know more about Al Qaeda's threat to AMC's West Coast strategic airlift assets than we do about the Cascadia Subduction Zone's earthquake/tsunami threat to McChord AFB. Seasonally, we have "Tornado Awareness Week" or "Hurricane Awareness Week," but there are more threats to the Air Force mission than just wind and wave. Moreover, there are no plans to relocate or reposition a base's primary mission (aircraft repair, avionics, etc.) if its real property is rendered unusable by a natural disaster, or if the cost to repair, replace, or reconstruct its mission-specific facilities is deemed too expensive.

A final factor: the continuing reduction of Air Force bases through the Base Realignment and Closure Commission (BRAC) process has resulted in a depletion of redundant facilities *i.e.,* similar repair facilities, hangers, civilian expertise, throughout the Air Force. Thus, a second order effect of BRAC makes each Air Force base, literally, its own "island" of expertise. If a base with a unique capability, say C-5 repair, is rendered inoperable following a natural disaster, there are few, if any, Air Force bases capable of continuing that particular mission without impairing their own unique work. In a zero-sum game, one-of-kind base functions put Air Force missions at risk.

Background. Many factors have been used to establish, modify, and drawdown bases in the last one hundred years. Flying weather, geography, topography, and even soil composition are a few of the factors that have played into these decisions. Coupled with the inevitable mixture of local, state, and national politics, the siting of Air Force bases has been an inexact, but enduring science in the last century. Interestingly, the current base laydown structure seems to validate Army Air Corps basing decisions made prior to and during World War II. In fact, the June 1963 the "Ideal Base Study," commissioned by then Secretary of Defense Robert S. McNamara, found that Air Force forces were "…accommodated at locations closely approximating the ideal."

The College of Aerospace Doctrine, Research and Education (CADRE) *Quick-Looks* are written by military defense analysts assigned to Air University (AU) and beyond are available electronically at the AU Research Web Site (https://research.maxwell.af.mil) or (http://research.airuniv.edu). Comments are encouraged. Send to: CADRE/AR, Director, Airpower Research Institute, 401 Chennault Circle, Maxwell AFB AL 36112-6428 or e-mail: cadre/arr@maxwell.af.mil.

Quick-Look 06-05

APRIL 2006

On the other hand, Nature has seen fit to destroy or severely damage more than one of these bases, highlighting the one factor missing from these basing deliberations: susceptibility to natural disasters.* The BRAC—now in its fourth round of closings—has ignored the possibility that some installations may a fall within a "tornado alley," lie on a flood plain, straddle a seismic fault, or operate adjacent to a potentially active volcano.

Recommendations. Just like terrorist vulnerability assessments, natural disaster risk assessments must be part of every Air Force base's planning. This is not to be confused with a "Hurrivac" plan—this must be a total assessment of the probabilities for natural disaster occurrence by type, followed by the creation of a set of SRR-like actions to mitigate loss of base functions. Local, state, and federal resources, historical climate/natural phenomena data, base histories, and national imaging resources should all be used to make a complete picture of potential threats. The Stafford Act (P.L. 93-288) requires the US Geologic Service (USGS) to provide notification, to include forecasts and warnings if possible, for earthquakes, volcanoes, and landslides. Its myriad of products should be a key player in any assessment process.

Some questions to be asked in any base natural disaster risk assessment:

— Are there other Air Force bases (to include Guard and Reserve installations) with similar missions that could host evacuated units for a protracted period of time without degrading their own efficiency?

— Are there ranges, ramp space, avionics shops, and hangers available for like weapon systems?

— Would overcrowding have a significant mission impact on the host and degrade its combat readiness?

— If so, are there sister service bases and facilities that could be used if Air Force facilities are not available? Are they compatible with Air Force weapon systems, to include logistics support and maintenance capabilities?

— Could Air Force installations closed/reduced by BRAC be reopened to accommodate long-term relocated missions *e.g.*, C-5 and B-52 maintenance at the former Kelly AFB?

— Are mechanisms in place to allow this without renegotiation/increased infrastructure costs?

— Can codicils be added to ongoing base closure negotiations that would allow their use in a national emergency without incurring major costs?

— Are prime weapon system contractors' facilities *e.g.*, General Dynamics in Ft. Worth or Boeing in Washington State, available for use?

— What are the secondary and tertiary order effects for any natural disaster? These could include, but are not limited to, a) disruption of lines of communication (roads closed, bridges washed out, microwave towers downed, etc.); b) displacement of the work force; and c) interruption of normal supplies and equipment in favor of relief efforts.

Such assessments should form the basis for every Base Natural Disaster Survival, Recovery and Reconstitution Plan. The result will assure a continuum of effort across the Air Force following any natural disaster. Knowing the threat and planning for all of its consequences are the keys to managing the risk.

* Clark AB, The Philippines (Mt. Pinatubo) and Homestead AFB, Florida (Hurricane Andrew) are examples of Nature's base closure policy, but these closure decisions were strongly influenced by the political wrangling that preceded both events. Negotiations with the Philippine government were nearing conclusion when the volcano erupted; Homestead AFB was on the BRAC short list when Andrew came ashore.

Quick-Look 06-05

April 2006

Addendum: "So, Who's Vulnerable?" A Non-Exhaustive List

Tsunamis: Much has been written in the past year about these devastating tidal waves, yet West Coast tsunami analysis indicates that **Vandenberg Air Force Base** is not considered susceptible to inundation. Neither is **McChord AFB** in Washington State. Likewise, the parallel runways at **Anderson AFB**, Guam are well over 500' MSL and likely in no danger from tsunamis. On the other hand, the main runway at **Hickam** AFB, Hawaii (8L/26R, AKA the "Reef Runway") is less 13' MSL. According to the Oahu Office of Civil Defense, it would be completely inundated, along with portions of the Hawaii Air National Guard complex to its west by any tsunami. Not only is this runway shared by the **Honolulu International Airport**, it is an alternate landing site for the **Space Shuttle**.

Floods: Obviously, any base that sits near or on a major body of water is susceptible to flooding. What is not widely recognized is that the Air Force has the safety responsibility for 23 dams located on Air Force bases in the CONUS. Given the destruction caused by breached levees in New Orleans and the recent collapse of earthen dams in Hawaii and Sacramento, California, dam breaches on Air Force installations must be factored into any natural disaster vulnerability assessment.

Tornados: Oklahoma is second only to Texas in the average number of tornados in the past 50 years (124–52), but has had more tornados (7.5 *vs.* 4.5) per 10,000 square miles than the Lone Star state. Given its relative size compared to Texas (268,601 square miles vs. 69,903 square miles), the tornado threat to Air Force installations in Oklahoma (**Vance**, **Tinker**) is exponentially greater than to the numerous Air Force installations scattered about Texas. In addition, factoring in the frequency of death, injury, and cost of damages, Oklahoma (4th) ranks well above Texas (17th) in tornado risk.

Earthquakes: More than just a California phenomenon, other sections of the USA are vulnerable to earthquakes. Research on the recently discovered Cacadia Subduction Zone off the Oregon/Washington coast places **McChord** AFB at risk. To the East, **Hill** AFB lies in the Wasatch Fault Zone. Further east, **Scott** AFB sits between two major geologic faults: the New Madrid Fault and the Wabash Valley Fault. Geologists predict a major earthquake (magnitude 7+) in that region within the next 50 years. **Little Rock** AFB, at the southern tip of the New Madrid Fault, is also at risk. On the East Coast, **Charleston** AFB lies in another geologic fault zone and is considered vulnerable. During the period 1974—2003, Alaska had 12,053 earthquakes of a magnitude 3.5 or greater—57.2% of all reported earthquakes in the United States. All Air Force bases in Alaska are at some risk.

Volcanoes: Alaska has 5 of the 18 most dangerous volcanoes in the United States. Luckily, these are located on islands in the Aleutian Chain and pose no threat to Air Force installations. On the other hand, although Mt. Rainer, southeast of **McChord** AFB, poses no immediate threat of *lahars* or lava flow to the facility, volcanic ash from any eruption threatens flight operations and has caused the base's aircraft to evacuate on prior occasions.

The major threat from active volcanoes is to aircraft flying in the vicinity of an eruption. In the past 30 years, over 100 incidents involving aircraft engine of airborne volcanic ash have occurred. 20 percent of these involved significant damage to engines. In 8 cases, in-flight engine failures occurred.

Hurricanes/Typhoons: The danger to coastal Air Force bases is well known and well documented. Statistics from 1851 through 2004 show 92 direct hits on the US mainland. Coincidentally, there have been 92 major hurricanes (Category 3-5 on the Saffir-Simpson scale) in the same time frame, out of a total of 273 recorded storms. Florida, with the most coast line, has had the most major storms (35), followed by Texas (19), Louisiana (18) and North Carolina (12). The Delmarva Peninsula has had only one major storm, while the Northeast Corridor of New York Connecticut, Rhode Island and Massachusetts have combined for a total of 15. Since hurricanes are—in effect—living, moving things—it is difficult to predict with any certainty where major storms will strike. Past climatology bears close scrutiny in developing any hurricane threat assessment.

Quick-Look 06-05

Strangers in a Strange Land:
The Federalist Papers, the Air National Guard and Homeland Defense

John L. Conway III

"The Army National Guard from Lafayette, Louisiana camped in the lobby of the hospital in full combat gear and loaded M-16s. It felt good to have the military come in. This will be the finest hour of the American military command, far exceeding their work for any war in foreign countries."

New Orleans Emergency Room Physician
5 September 2005[1]

Issue:
The American people expect their military to respond to any national disaster, man-made or natural. And nobody in crisis cares who they are or what branch they represent. No one clinging to a rooftop in New Orleans' Lower Ninth Ward questioned the Title 32 or Title 10 status of their rescuers and no military medical caregiver will have to prove he or she is on active duty to provide aid and comfort to their patients. The only "patches" they will care about are those that bind their wounds. This is the immediate reality of disaster.

Nevertheless, the military is constrained from providing immediate disaster relief response except in the most extreme situations.[2] Moreover, the military is not designed to be a first responder, nor does the *National Response Plan*, nor the newer *National Response Framework*, envision it in that that role. Use of the military is at the far end of a wide range of options available to disaster response planners.

While all of the Services have performed admirably in relief operations, Americans have come to view the National Guard as the military's "first first-responder." Its response to Hurricane Katrina and its continued relief roles in smaller natural disasters have been widely-chronicled success stories. However, it's dramatically increased overseas role since 9/11 created worries that it will be unable to respond effectively at home for lack of personnel and equipment. Calls for a decreased overseas Guard presence continue as states realize that their "first first-responders" may be needed at home sooner rather than later.

The Air Force embraces the Air National Guard as a full partner in the Total Force and relies on it for combat missions every day. In turn, the Air Guard is justifiably proud to be a full partner and is loath to give up this hard-earned status to become a standby force for natural and man-made disasters instead.

Should the Guard reduce its overseas presence in favor of an increased Homeland Security role? If so, which new roles should it undertake and which current roles should it abandon?

---

[1] Quoted in the *Decatur* (AL) *Daily*, 5 September, 2005.
[2] Commanders always have the prerogative to conduct immediate response operations near their installations to protect life and property, and to mitigate human suffering.

Air Education and Training Command — "Develop America's Airmen Today . . . for Tomorrow"
Air University (AU) *Quick Looks* are written to generate discussion on topics of interest to the larger national security community. These Quick Looks are available electronically at the AU Research Web Site (https://research.maxwell.af.mil) or (http://research.airuniv.edu). Comments are encouraged. Send to: AFRI, Director, Air Force Research Institute, 155 N. Twining, Bldg 693, Maxwell AFB AL 36112-6428 or https://www.afresearch.org/skins/rims/home.aspx?
AU Quick Look 08-08

Background:
Military response to CONUS disasters is a "pull" system vice a "push" one: local, state, and tribal authorities, the real "first responders," must be overwhelmed and formally ask for help from the federal government before it can be sent. Even then, this may not include the DOD. Before federal military aid is requested, the governors of the several states and territories can call up the National Guard. Further, they can ask for National Guard personnel from other states to assist them under Emergency Management Assistance Compacts (EMACs)–all under the requesting state's control. In this way the Guard is able to respond well ahead of any federal military effort. Moreover, Guard forces in State Active Duty status or under Title 32 status have the freedom of action to restore public order (read: no *Posse Comitatus* restrictions) that federal forces do not.[3]

If federal (Title 10) forces are needed, an elaborate process must take place in order to get them on the ground. According to the White House analysis of the military response to Hurricane Katrina, DOD assets were dispatched only after an approval process requiring *twenty-one* separate steps. Lag time between the request and federal "boots on the ground" was measured in days, not hours.

When the time between "search and rescue" and "recovery" operations is usually less than 72 hours, summoning the most available forces makes eminent sense. Given its organization, resources, and local basing (units in over 2000 communities), the military first-response mission belongs to the National Guard, not active duty forces.

The issue, then, is not who, but how?

Currently there are no standing requirements for any military organization—much less the Guard—to organize, train, and equip for Homeland Defense. The 2008 National Defense Authorization Act requires the Department of Homeland Security and the DOD to state requirements, but with only one National Guard representative assigned at DHS, this will be a long process before tangible results will be seen.

Way Ahead:
In the absence of any planning guidance and in view of the uncertain nature and timing of natural and man-made disasters, the National Guard Bureau and the Adjutants Generals of the several states must create response packages from their existing resources. In time, downward-directed guidance will be provided, but an opportunity exists for the Guard to shape that guidance in reality today, whereas it might have to accept someone else's guidance in the future.

Due to its sheer size, the Army National Guard would obviously to be the major responder, but there are important roles for the Air National Guard beyond Search and Rescue.

The new Joint Cargo Aircraft (JCA) will be a perfect short haul platform for in-close airlift support on damaged or unimproved airfields. Air Guard units could lift emergency personnel or Chemical, Biological, Radiological and Nuclear (CBRN) response units close to the disaster site and evacuate victims as well. Air Guard units with the JCA would have a dual Designed Operational Capability (DOC) to reflect an airlift capability for Irregular

[3] State Active Duty (SAD) is paid by state funds. In contrast, in Title 32 status (Title 32, USC 902(a) "…the Secretary of Defense may provide funds to a Governor to employ National Guard units or members to conduct homeland defense activities that the Secretary determines to be necessary and appropriate for participation by the National Guard units or members, as the case may be."
In both instances, National Guard personnel are under the command and control of the Governor through the state Adjutant General.

AU Quick Look 08-08

Warfare (IW) and Major Combat Operations (MCO), and a secondary DOC for Disaster Relief. A dual DOC would allow training dollars for both missions using the same equipment.

Medical personnel are obvious dual-capable Air Guard assets, particularly MEDEVAC units. Here again, medical units would require modification to their existing DOC statements for disaster relief missions. With a disaster response DOC, there would be additions to the units' Unit Manning Document (UMD) to provide for unique first-response medical needs beyond the typical emergency medical specialties. For example, there is a new physician certification in disaster medicine that encompasses all phases of disaster medical delivery, which is a natural complement to existing unit medical capabilities.[4] Veterinarians will also be needed to care for abandoned livestock and pets, and to assess public health concerns.

Another atypical military medical specialty to be added is gerontology: the aftermath of Hurricane Katrina saw large numbers of nursing home patients literally abandoned by their long-term care facilities and seeking assistance. This will strain triage of disaster victims, but their care and processing onward must be planned for and addressed in an overall disaster response package. Psychiatric specialists will also be required to address Post Traumatic Stress Disorder (PTSD) patients as well as "the worried well;" those without apparent injuries who nevertheless seek medical assistance.[5]

States should use the long-standing imagery interpretation skills in the Air National Guard, assigning Air Guard imagery experts to state emergency staffs during a crisis. It is one thing to look at imagery; it is quite another to understand or make proper use of it. During Katrina relief operations, over 690,000 square kilometers of imagery was sent to state and local decision makers, yet it is unclear how if it was even used. Up-channel requests for imagery support were poorly articulated by civilian decision-makers who were unfamiliar with imagery capabilities. Air Guard imagery experts would be invaluable for assessment.[6]

A post-Katrina legal review by the American Bar Association outlined numerous unique legal issues that arose during Katrina relief operations and pointed up the need for on-scene legal assistance in disaster response.[7] Again, in the absence of downward direction, Air Guard units should seek to increase their unit JAG manning to prepare for legal needs during support to disaster operations.

The American public's historic wariness regarding a standing army and its belief in the militia system of defense makes the use of the Guard for disaster relief far less contentious than the imposition of federal forces. Americans see "the" Army and "their" Guard quite differently. Frequent federal response to disasters in "their" Homeland will be met with increasing skepticism which, in time, could spawn outright mistrust.

Defense of the Homeland, then, is primarily the Guard's mission. In the planning vacuum that currently exists, it is the Guard that must take the lead.

[4] Homeland Security Presidential Directive 21, 18 October 2007. See also the American Board of Physician Specialties website listing: "American Board of Disaster Medicine Certification Exams."
[5] *The Worried Well*, USAF Institute for National Security Studies (INSS) Occasional Paper Number 53, April 2004.
[6] UAV assets could also provide assessment, but varying interpretations of the law make this a murky area *i.e.* "surveillance" of Americans is legal by the Guard in State Active Duty or Title 32 status, but illegal by Title 10 federal agencies under the *Posse Comitatus Act*. (18 USC §1385) "Reconnaissance" by federal agencies is legal, but the difference is not apparent without a scrupulous legal review.
[7] "Hurricane Katrina Task Force Subcommittee Report," *ABA Standing Committee on Law and National Security*, February, 2006.

AU Quick Look 08-08

"The perpetual menacings of danger oblige the government to be always prepared to repel it; its armies must be numerous enough for instant defense. The continual necessity for their services enhances the importance of the soldier, and proportionally degrades the condition of the citizen. The military state becomes elevated above the civil. The inhabitants of territories, often the theatre of war, are unavoidably subjected to frequent infringements on their rights, which serve to weaken their sense of those rights; and by degrees the people are brought to consider the soldiery not only as their protectors, but their superiors."

Federalist No. 8, November 20, 1787

Air Force Research Institute

AU Quick Look 08-08

Montag Speaks Farsi:
A New Approach to Air Force Language Needs

John L. Conway III

Issue: The 2005 release of the DOD Language Transformation Roadmap heralded a new beginning for language education in DOD.

Computer courses in language were added at ACSC and Air War College, and the Air Force Academy added two new language offerings and upped its language requirements for non-technical majors.

Foreign Language Proficiency Pay (FLPP) was significantly raised; the Guard and Reserve finally received FLPP parity; and even qualified non-intelligence Air Force civilians were offered FLPP, albeit at an inexplicably lower rate.

Air University established a Center for Language and Culture; Air Force ROTC created more opportunities for language immersion for its cadets; and the Air Staff moved the language office to the A1 community.

So, are we there yet?

Sadly, no.

None of these actions—none—have resolved the issue of language proficiency for the Air Force. Except for the cryptoanalytic career field, no Air Force language requirements exist. The Air Force still does not know if it has too many or too few language capable personnel: "mandatory" surveys of language skills have stalled at about the 50 percent return rate.

Opportunities for officers to use their language skills remain reserved for the International Affairs Specialist Programs or for Attachés (read; Majors and above), so Captains and Lieutenants cannot expect to use their language skills for the first eight years of their careers.

Three years after the publication of The Roadmap, how does the Air Force insure it has the necessary language abilities, in the right numbers, to meet its needs?

Background:
We are an International Civil Aviation Organization (ICAO) Air Force. English is the standard language, with very few exceptions, in international airway communications, so why should we need another? Despite its "melting pot" image, America is a big English-speaking island: rarely do Americans encounter another language—other than Spanish—in their normal day-to-day environment. We are not Europeans, where other

Air Education and Training Command — "Develop America's Airmen Today . . . for Tomorrow"
Air University (AU) *Quick Looks* are written to generate discussion on topics of interest to the larger national security community. These Quick Looks are available electronically at the AU Research Web Site. Comments are encouraged. Send to: AFRI, Director, Air Force Research Institute, 155 N. Twining, Bldg 693, Maxwell AFB AL 36112-6428 or https://www.afresearch.org/skins/rims/home.aspx?

AU Quick Look 08-07

countries, cultures, and languages are just a three hour (two hours if you're on the Autobahn) drive away. In the Americas, you can travel from Tierra del Fuego to Nome and need only four languages to communicate; three if you skip South Louisiana and Quebec; only two if you also bypass Brazil.

The American educational system is not producing many language scholars either. Despite repeated calls for language education in elementary school, few courses exist. High school curriculums rarely include mandatory language requirements even for College Preparatory tracks. Relying on U.S. colleges and universities to produce the language scholars is not a panacea: about half of those that host Air Force ROTC units don't offer much more than the standard fare of French, German, and Spanish.

The International Center for Language Studies calculates that it takes 150 hours of classroom instruction (plus 2-3 hours of preparation per each five hours in class) to achieve a Defense Language Proficiency Test (DLPT) score of 1/1 in the Romance and Germanic languages, considered the easiest to master. To meet DOD's DLPT goal of 3/3, quadruple that number and add a few more hours of coursework; for Arabic, Mandarin Chinese, Japanese and Korean, considered the hardest, triple *that* number. That's a lot of college coursework.

The Defense Language Institute - Foreign Language Center, at Monterey could handle more students, but its primary mission is to train cryptoanalytic linguists, and it would be hard pressed to expand much further.

Way Ahead:
Given all of the above, there is no current method by which the Air Force can predict its needs, much less meet them. In the absence of a better language acquisition method, here are two alternative approaches worth further debate.

The "Every Marine a Rifleman" Solution:
Simply stated, all incoming Air Force officers must present proficiency in a language in order to be commissioned. This will ensure language knowledge throughout the officer corps, albeit at its lowest tier and with varying degrees of competence However, it would negatively impact the intake of college graduates in the science and engineering fields. Given the technical bent of the Air Force, this is perhaps a self-limiting condition.

The "Fahrenheit 451" Solution:
This approach takes its name from the Ray Bradbury science fiction novel. In the future painted by Bradbury, books are outlawed but knowledge within them is saved was by individuals literally becoming the books themselves. At the end of the book, its protagonist, Montag, meets a group of men who have memorized books and he vows to become one of them; to become a book himself.

Since the intelligence community has recently considered requiring all intelligence officers to have language skills, it would be an excellent cohort to implement this approach; each intelligence officer would become a language.

To produce enough linguists in a variety of languages, each newly minted intelligence officer would be assigned a language from DOD's Strategic Language List.[1] For the rest of his/her career (including Air National Guard or Air Force Reserve duty), it would be his/her responsibility to maintain proficiency in it, even if his/her career never requires its use. In this way, Lt Jones would become "Mandarin Chinese" and Lt Smith would be "Arabic." If needed and where needed, these officers would be trained and available.

[1] Either a Stronghold or an Investment Language.

AU Quick Look 08-07

To do this successfully, however, requires some practical guidance.

-Don't continue to insist on mid-career initial language training, even as some sort of "language appreciation" initiative. Start at the post-commissioning phase. It will take longer to have an impact, but it will produce better results than a headlong rush to the language lab by everybody in the Air Force.

-To add rigor to this program, upgrades for the Intelligence Specialty Badge would be tied to language proficiency *e.g.* make a DLPT score of 2/2 a part of the criteria to award the Master designation.

-As a further incentive—and to offset costs for tutors, tuition, or travel—make knowledge of a foreign language a prerequisite to the award of the intelligence AFSC (*i.e.* "No language, no AFSC"); this will allow all intelligence officers with language proficiency to be eligible for Foreign Language Proficiency Pay (FLPP).

-If a language major or native speaker becomes an intelligence officer, the language concerned would be his/her language skill and proficiency would be validated by the DLPT.[2]

Successful acquisition of language skills is measured in years; maintaining a language skill requires a lifetime. Air Force personnel must be able to take advantage of refresher training courses, periodic immersion classes, and distance learning programs to maintain proficiency.

"Would you like, some day, Montag, to read Plato's *Republic*?"
"Of course!"
"*I* am Plato's Republic. Like to read Marcus Aurelius? Mr. Simmons is Marcus."
Ray Bradbury, *Fahrenheit 451*

Air Force Research Institute

[2] Recruiting native speakers as a quick fix may not work. Clearance issues may be difficult to overcome for some first generation Americans. In addition, emerging scholarly research that indicates a substantial decline in a first generation native speakers' ability in their native tongues as they try to assimilate into mainstream America.

AU Quick Look 08-07

In Service to the Nation

An Air Force Strategy for 2018-2023

Background: "In June 1990, the Secretary of the Air Force, Donald B. Rice, issued a Department of the Air Force White Paper entitled *The Air Force and U.S. National Security: Global Reach—Global Power* which set the stage for the evolving role of the Air Force as a prime instrument of national security policy and strategy."[1] This widely acclaimed white paper helped transition the Air Force and the nation from a military mindset of forward defense, born in the Cold War, to a more flexible and less urgent posture predicated on our ability to move forces and employ force anywhere on the globe, as needed.[2] It was a very powerful and enduring message for the Air Force.

Capitalizing on this success, Air Force leaders added "Global Presence" (1995), "Global Engagement" (1997) and "Global Vigilance, Reach, Power" (2000).[3] However, by 2000, the larger message encapsulated in the connected phrase Global Reach—Global Power had been substantially lost. Global Reach became the advocacy nameplate for Air Mobility Command; Air Combat Command got the Global Power brand. "Stove piping" the strategic message into specific platform-centric or programmatic statements isolated other Air Force operational and support elements. Air Force "strategy" was tied tightly to the POM.

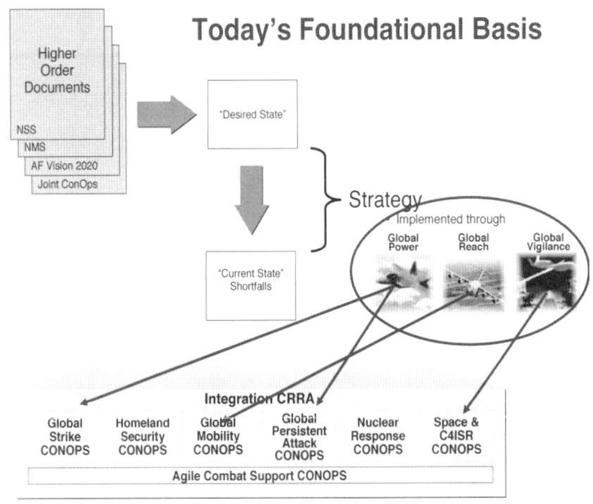

Figure 1: Separate strategy documents have evolved largely into advocacy statements for specific functions. An overarching framework is missing.

Issue: Global Reach, Global Power and Global Vigilance, as defined and used today, fall short of the full range of strategic options that the United States Air Force can bring to the fight. Normal bureaucratic momentum has been to develop

[1] Moorhead, Glen W., III, *Global Reach-Global Power and the USAF Tactical Air Forces*, 15 APR 1991, DTIC Number: ADA237060
[2] The Center for Naval Analysis cites this AF white paper, among other factors, as a shaping influence for the construct of Navy strategy in the 1990's. *U.S. Navy Capstone Strategies & Concepts (1970-2007)*-CNA briefing Sep 2007
[3] Air Force Magazine online: http://dc01-cdh-afa03.tranguard.net/AFA/DocumentFile/afvision/

Air Education and Training Command — "Develop America's Airmen Today . . . for Tomorrow"
Air University (AU) *Quick Looks* are written to generate discussion on topics of interest to the larger national security community. These Quick Looks are available electronically at the AU Research Web Site (https://research.maxwell.af.mil) or (http://research.airuniv.edu). Comments are encouraged. Send to: Director, Air Force Research Institute, 155 N. Twining, Bldg 693, Maxwell AFB AL 36112-6428 or e-mail: AFRI.AFRResearchers@maxwell.af.mil.

AU Quick Look 08-04

separate, specialized "homes" for each new capability or splinter group when interdependence *within* the Air Force is critical to effective articulation of both strategy and advocacy. In use, the separate AF strategy documents have largely evolved into advocacy statements for specific functions within the framework of AF contributions to national security. As such, an overarching, structured AF strategy is missing.

Way Ahead: In its simplest form, strategy is about ends, ways and means. As a service, AF has focused (perhaps over-focused) on the "means," that is, the people, systems and sustainment that can be "purchased" through POM actions. However, as an organization that is "effects-based," we need to rebalance these efforts starting with the *ends* of strategy. For the Air Force as a service in support of the National Military Strategy, these ends are Global Reach, Global Power and Global Vigilance. The definitions that follow are not radically new. Rather, each is "recast" to be more inclusive of all AF elements and the three principal domains in which we operate, as follows:

Global Reach is operational access—it is connectivity to the objective through air, space and cyberspace. Global Reach takes varied forms dependent upon circumstances, for instance, it could be a space-based line-of-communication, an air route for a C-17, or a portal-to-portal Internet connection. Global Reach is interdependent with the other pillars. For example, Airmen must first secure control of (or the necessary degree of superiority over) its principal domains (air, space and cyberspace) in order to exercise Global Reach.

Global Power is the ability to create and sustain effects through air, space and cyberspace. These effects encompass a full range of kinetic and non-kinetic, lethal and non-lethal, constructive and destructive options prosecuted through air, space and cyberspace either individually, or more likely, in a synergistic, mutually supporting campaign. Cross-domain integration in air, space and cyberspace enables greater speed, precision and reliability than options restricted to a single domain. This synergy produces the capability for more discrete effects with proportionally less collateral damage and fewer counter-productive effects. Cross-domain dominance enhances the Air Force's ability to provide sovereign options in all five domains: air, space, cyberspace, land and sea.

Global Vigilance is the situational awareness required for understanding both the necessity for action and the character of the effects necessary to achieve a revised condition or end state. It is the underpinning element, the foundational starting point for Global Reach and Global Power. Global Vigilance includes both the collection capabilities and the cognitive processes necessary to create that situational awareness. It identifies linkages and assesses the potential for success. Further, it validates actual successes or progress achieved through the application of Global Power. Global Vigilance is dependent upon Global Reach, that is, routine access through air, space and cyberspace to gather data and disseminate intelligence.

There are numerous examples of how the AF strategy can be realized but one of the great enablers, or *ways*, to achieve these AF strategic ends is through strategic partnering. These partnerships form from collaborative relationships built on common goals that enhance the United States' ability to achieve objectives while simultaneously aiding our partners to achieve theirs. It includes a broad range of activities with both state and non-state entities and may effectively act or be formally recognized as a coalition or alliance. It may include traditional mil-to-mil activities like foreign internal defense or international military training and education programs. It could also be cooperative engagement with non-governmental organizations during a humanitarian assistance mission or with an academic institution working to increase understanding in a social field or new findings in a technical sense. This burden-sharing, across a nearly unlimited array of partners, increases the opportunity for achieving US objectives while concurrently decreasing risk.

Traditional values of the Airmen's art (like speed, range and precision) are frames of reference that transfer easily between air, space and cyberspace. These domains, because of their near-boundless natures, also foster the Airman's global perspective—a natural extension of the airpower legacy (from providing vantage just beyond the nearest hill to transiting continents, seas, space, and cyberspace.) The Air Force uses these Air Force common traits in the global commons of air, space and cyberspace to produce these strategic ends for protecting the nation and securing its interests: ***Global Reach***, ***Global Power*** and ***Global Vigilance.***

Air Force Research Institute

Disclaimer: Opinions, conclusions, and recommendations expressed or implied within are solely those of the author and do not necessarily represent the views of the AU, AFRI, the United States Air Force, the Department of Defense, or any other US Government agency. Cleared for public release: distribution unlimited.

AU Quick Look 08-04

Operationally Responsive Space

Dr. Dale L. Hayden

Issue: The US has a proven ability to launch highly capable, highly sophisticated and expensive satellites into orbit. Further, the US's great reliance on space assets has created a potential vulnerability out of its dependence. The former Soviet Union, and most recently China demonstrated anti-satellite (ASAT) capabilities at a time when US space control and defensive space operations remain problematic.

A change in philosophy and approach is needed for space operations to be responsive to the warfighter and reduce the strategic vulnerability. The history of the US space effort shows a reliance on medium and heavy lift vehicles, boosting even more expensive satellites into orbit. In an era of constrained resources and with an increasing demand for space assets, the Air Force must fully embrace the concept of Operationally Responsive Space (ORS). Former Under Secretary of the Air Force, Peter B. Teets, defined ORS in testimony before Congress as, "a more responsive, reliable, and affordable lift family capable of fulfilling both current and future launch requirements, and the corresponding responsive and affordable satellites."

Background: Central to national security and meeting the warfighter's needs is assured access to space through reduced systems cost accompanied by the ability to launch and replenish assets when needed. The US space effort does not have a history of timely response to the warfighter. For example when it became clear in September 1990 during the planning stages of Desert Storm that the coalition did not have adequate communications capability, senior leadership approved the launch of an additional Defense Satellite Communication Systems (DSCS). This DSCS satellite reached orbit February 1992, over a year after hostilities ended.

While the nation has made strides over the past decade, significant challenges lie ahead. Although two new families of spacelift vehicles now exist -- the Evolved Expendable Launch Vehicle (EELV) program composed of Boeing's Delta IV family and Lockheed Martin's Atlas V family of medium and heavy lifters—neither significantly reduce launch costs. In 2006, the National Security Space Launch Requirements Panel concluded that while the Delta IV and Atlas V can meet the nation's space launch needs through 2020 (the end of their study period), they were "largely uncompetitive in today's commercial market," and that because ORS operational concepts were in the formative stages, it was premature to specify launcher requirements. Launch costs today generally run about $10,000 per pound for low earth orbit, resulting in launch cost of from $500 million to $1 and $2 billion per effort. Commercial companies, like *SpaceX*, are testing lifters for operational use that range in the area of $7 million per launch, but none has yet successfully placed a satellite in orbit.

Way Ahead: In May 2007, the Air Force took a major step in the right direction when it opened an ORS Office at Kirtland AFB with the intent to focus on smaller satellites, smaller boosters and getting those capabilities into the hands of the warfighter. The next step should be a second *Space Launch Modernization Study* that brings the requirements for ORS together and focuses on light to medium vice medium to heavy lift. The first study ultimately lead to EELV, moving America away from legacy Intercontinental Ballistic Missile (ICBM) based launch systems to modern, highly reliable vehicles. The new study must revolutionize light spacelift, much as the first did for medium and heavy lift with the objective of dramatically lowering spacelift cost with an outcome where the measure of merit is launches per day and month vice launches per year.

Air Education and Training Command — "Develop America's Airmen Today . . . for Tomorrow"

Air University (AU) *Quick Looks* are written to generate discussion on topics of interest to the larger national security community. These Quick Looks are available electronically at the AU Research Web Site (https://research.maxwell.af.mil) or (http://research.airuniv.edu). Comments are encouraged. Send to: AFDDEC/AR, Director, Air Force Research Institute, 155 N. Twining, Bldg 693, Maxwell AFB AL 36112-6428 or e-mail: dale.hayden.ctr@maxwell.af.mil

AU Quick Look 08-02

A smaller, less expensive lift vehicle would be accompanied by smaller, less expensive satellites. The argument of small versus large satellites is not new. It is a discussion going back decades inside and outside of the Air Force. The nation will always need large, complex and correspondingly more expensive satellites, especially in geostationary orbit. However, the US needs a strategy that includes a mix of large and small satellites to increase capability and simultaneously reduce vulnerability if our future success as a space faring nation is to be guaranteed. Smaller satellites need not be viewed as less capable. The same revolution that took place with personal computers is also available to space operations. Further, increasing satellite numbers that might have shorter duration missions need not be viewed as a limitation. Much like aircraft acquisition, increasing the number of satellites and launchers coming off an assembly line will reduce cost per unit. Lower cost and decreased lifecycles can take better advantage of the latest technologies that dramatically increase capability verses size. Readily available satellites, either spares on orbit or in warehouses, could also reduce a constellation's vulnerability to attack. If the US had the ability to replenish a lost asset within hours or days, it reduces the likelihood that an adversary would attempt to destroy it; simply why bother.

Additionally, the Air Force needs to look at patterning with commercial industry to place production and assembly sites for satellites and launch vehicles in close proximity to the launch ranges at Vandenberg AFB and Patrick AFB. Production and warehousing on-site will further reduce life cycle costs and dramatically reduce launch response times. Challenges do exist to this strategy, but senior leaders must make the right choices to compete against conglomerates, like the European Space Agency and China where legacy systems do not play a role and launch facilities are near or on the equator, where less lift is required to get a satellite in orbit.

Operationally Responsive Space provides the ability for persistent global reach and vigilance through enhanced ISR, making possible a revisit rate sufficient to hold targets at risk. It also augments global power, in any orbital plane through redundancy, with the prospect of reconstitution to defend on-orbit and support assets. Simply stated, the key to ORS is responsiveness to the warfighter by providing options and alternative strategies.

Operationally Responsive Space, if fully embraced by the Air Force and private industry, will transform space operations for the 21st Century.

Air Force Doctrine and Development Center
Air Force Research Institute

AU Quick Look 08-02

Avoiding Weaponizing Space

Dr. Dale L. Hayden

Issue: The overarching issue of future space control surrounds the issue of space warfare and its corollary weaponizing space. Space as the ultimate high ground remains the one medium, unlike ground, air and sea, where weaponization remains an issue. In one aspect, space was weaponized decades ago. Space ISR assets have long gathered data for use by combat forces. The former Soviet Union, and more recently China successfully demonstrated anti-satellite weapons (ASAT) capabilities. The concern, then, is not weaponization, but rather weapons positioned on-orbit in space.

Background: International agreements do not prohibit space weaponization. Treaties and conventions taken together prohibit placing nuclear and other weapons of mass destruction in orbit around the earth or the moon, and prohibit placing military installations or weapons on the moon or other celestial bodies. They also declare that space is to be used exclusively for "peaceful purposes." Pressures in the next 15 years will bring additional emphasis to the issue of weapons in space. Some argue that the US has already taken the first steps by pulling out of the 1972 ABM Treaty in June 2002. The US dependence on space and the necessity to protect its assets coupled with technological evolution will make placing weapons in space viable. In addition, at some future point weapons positioned in orbit will be able to hit ground, sea or airborne targets.

In the future, the US may face a dramatic, potentially catastrophic event in space similar to the Soviet Union placing nuclear missiles in Cuba during the early 1960's. President John Kennedy believed if he had not acted to remove the missiles the American public would have called for his impeachment. A president in this century may well face similar pressures where the US finds itself at risk due to an advisory placing weapons on-orbit. If the US does not take positive steps to avoid weaponization, then the pressures may become overwhelming. What, then, might be viable alternative strategies to avoid on-orbit weapons?

Way Ahead: Numerous strategies exist that can be implemented that avoid the necessity of placing weapons on-orbit. The most attractive strategy is global partnering. Through global partnering the international community can develop a system of relationships in space that encourages beneficial and benign behavior while still containing threats. Through economic and technical cooperation nations become interdependent and therefore, much less likely to act against their own interests. We already find ourselves partnering with other space-faring nations in such areas as the International Space Station. Both Russia and China have launched satellites for US based corporations. Additionally, the US and Europe are cooperating to avoid frequency overlap in the deployment of Galileo, a European version of Global Positioning System (GPS).

Global Partnering also lays the foundation for international negotiation, regulation and governance by the rule of law. Building upon such treaties as the 1967 Outer Space Treaty and associated legal regimes that dictate space activities is a viable strategy to enhance space control. Internationally, the US is party to a series of regulations in other medium that govern everything from air traffic control to commerce on the high seas.

Air Education and Training Command — "Develop America's Airmen Today . . . for Tomorrow"

Air University (AU) *Quick Looks* are written to generate discussion on topics of interest to the larger national security community. These Quick Looks are available electronically at the AU Research Web Site (https://research.maxwell.af.mil) or (http://research.airuniv.edu). Comments are encouraged. Send to: AFDDEC/AR, Director, Air Force Research Institute, 155 N. Twining, Bldg 693, Maxwell AFB AL 36112-6428 or e-mail: dale.hayden.ctr@maxwell.af.mil

AU Quick Look 08-03

Precedent exists to limit space activity through international negotiation and regulation. Following a US space based nuclear detonation during the early 1960s, the international community, to include the US, moved to ban weapons of mass destruction in space. While the blast was highly effective in destroying potential enemy space based assets, it also had an unintended consequence of disabling distant US satellites through electro-magnetic pulse (EMP). Building upon such precedent, agreements could call for the elimination of ASATs which many nations, including China, might well find attractive. The Chinese discovered a consequence of their successful ASAT launch in 2007 was a massive debris field that extends from less than 125 miles to more than 2,292 miles altitude; this range encompasses low earth orbit. The resulting clutter will affect all nations, including China, who desire to place satellites in low earth orbit. International negotiation and regulation is not a panacea, as this approach has difficulty influencing non-state actors and enforcement is often problematic.

Another means to protect space assets is to reduce the importance of any single satellite or group of satellites. This can be accomplished by having spares on-orbit or the ability to rapidly replace lost assets. If an advisory negates the use of one or more of our satellites, being able to rapidly reconstitute the constellation dramatically reduces our vulnerability. Rapid reconstitution is dependent upon full implementation of operational responsive space (ORS) where launch and satellites costs are reduced over today's standards. Responsive launch is required to replace satellites in a timely fashion and a reduction in satellite cost is essential to allow warehoused spares. This purely defense approach essentially makes any one part more resistant to attack and ultimately is less expensive and less controversial than placing weapons in space.

Another viable defensive strategy involves a more passive approach. Placing bodyguards, such as space mines, near a high-value asset could defend against an ASAT, but this strategy could produce an on-orbit debris field, an effect best avoided when possible. On-orbit dummy or drone satellites, sending out signals similar to the actual assets would provide protection and reduce vulnerability by confusing and overwhelming an adversary. Additionally, satellite hardening through redundant systems or making systems more impervious to attack provides little rational for an advisory to take the first steps in placing weapons in space. A problem with this approach is it could increase a satellite's vulnerability by creating new high value target through complexity, size and increased cost.

The Air Force must be ready when called upon. If that means protecting vital national interests through weaponizing space, then it must be prepared to do so. However, on-orbit weapons are not inevitable. The Air Force can take measures today that make placing weapons on-orbit unnecessary or at worst unlikely.

Air Force Doctrine and Development Center
Air Force Research Institute

AU Quick Look 08-03

ICBM Future Strategy

Dr. Dale L. Hayden

Issue: The nation has an aging intercontinental ballistic missile (ICBM) force providing national strategic defense. At its height, the US had 1500 land-based ICBMs. When the 564th Missile Squadron at Malmstrom AFB, MT closes in August 2008, that number will decrease to 450. Further, the US has not field-tested a nuclear weapon since 1992 other than through computer simulation. During the Cold War, nuclear deterrence via the triad provided the overarching umbrella to national security that no other weapon system could. However, with a reduced nuclear threat and a rapidly aging fleet, the question remains: should the nation relinquish the ICBM force and rely on its strategic bombers and ballistic missile submarines for nuclear defense or should it fund modernization of the triad?

Background: The reduction in missiles and that of the overall nuclear arsenal is made possible due to more stable international relations. However, Russia retains a significant nuclear force and China continues to add to its arsenal. Furthermore, the nuclear posturing of nations like North Korea and Iran serve to continue to make nuclear deterrence relevant. Ongoing modernization and upgrade programs of the fleet will allow the existing launchers, warheads and command and control systems to remain operational into the 2020 timeframe. However, mounting pressures will call into question the viability and necessity of maintaining land-based ICBMs much beyond the 2020 timeframe.

The launch control centers (LCC) housing the crews and the accompanying missile launch facilities (LF) were built and placed underground during the 1960's and are now nearing the 50 year mark. Although made of steel-reinforced concrete, they have not been survivable from direct nuclear attack for decades. In addition, environmental pressures, such as the water table, underground streams and shifting soils continue to slowly degrade the structures.

Though the triad served the nation well for over 60 years, public support for nuclear modernization is problematic. Telling is a survey of active duty Air Force officers where only 2 percent of the respondents believe the US needs a new generation of nuclear weapons. ICBMs have little or no play in the pressing topics of the day: Global war on Terrorism (GWOT) and irregular warfare. Further, eliminating one leg of the triad and moving to a bi-ad structure saves money at a time when the defense budget will most likely be significantly reduced over that of today.

The rationale for a nuclear triad is based upon a balance of deterrence and risk. Relying upon a nuclear bi-ad may well provide deterrence, but at an increased risk over that of a triad. The Defense Science Board Task Force on Nuclear Deterrence concluded, "Significant numbers of ICBMs deny any adversary the benefit of a limited attack. Without the ICBMs, surprise attacks against a handful of bomber bases and SSBN facilities, with plausible deniability, could drastically alter the correlation of forces." Simply put, how much risk is the nation willing to accept to guarantee its survival? If decision-makers do not want to substantially increase the risk to national survival, then continued reliance on the triad is a necessity and ICBM modernization is a must. Recognizing the political climate, any modernization effort must be affordable and be viewed by the American public and international community as a stabilizing factor while not tipping the balance of power or serving as a destabilizing factor.

Air Education and Training Command — "Develop America's Airmen Today . . . for Tomorrow"
Air University (AU) *Quick Looks* are written to generate discussion on topics of interest to the larger national security community. These Quick Looks are available electronically at the AU Research Web Site (https://research.maxwell.af.mil) or (http://www.afresearch.org). Comments are encouraged. Send to: AFDDEC/AR, Director, Air Force Research Institute, 155 N. Twining, Bldg 693, Maxwell AFB AL 36112-6428.

AU Quick Look 08-04

Way Ahead: The first step in creating a sustainable system is a modern command and control structure. To address the aging, non-survivable LCCs, the Air Force should move the equipment and crews above ground. This move can reduce life cycle costs–e.g., ease of maintenance–while taking advantage of the latest technologies. This strategy follows the precedent set in closing the Cheyenne Mountain Complex and moving its command and control functions to surface facilities. Vulnerability increases somewhat-noting they have not been survival from nuclear attack for decades–as they now become a softer target, but through redundancy much of the vulnerability can be mitigated. Placing the new command and control structure within the existing above ground missile alert facilities would also mitigate the vulnerability, as they are fenced and can be guarded. The next step in creating a sustainable force is equally important, modernization of the launch system.

The 450 Minuteman III launchers will exceed their life expectancy soon after the 2020/2025 timeframe. At that point, the nation will be faced with the decision to retire the fleet. Any follow-on system must be also affordable and sustainable. Both attributes can be addressed through use of a common launcher, either commercial or one meant for wider military application. A missile with wider application allows R&D cost to be spread over a larger number of vehicles, while at the same time reducing maintenance costs through using common hardware. The new launchers should be based above ground, again as with the LCCs, taking advantage of the latest technologies and reducing maintenance complexity and costs.

System affordability must also be addressed by reduced fleet numbers. Without a detailed analysis, it is difficult to determine the proper number of missiles. Four-hundred and fifty may or may not be the right number, but how far the nation should go is best determined by a requirements study that balances threat against risk against funding. Basing should be distributed and can use the existing wing structure at Malmstrom, F.E. Warren and/or Minot AFB, further enhancing survivability while keeping system cost in mind.

Improved command and control and a reduced number of new, single warhead ICBMs can be perceived as a stabilizing factor. Mobile launchers might increase survivability, but they could also serve as a destabilizing factor and increase cost, therefore static launchers are preferable as a future strategy. If a study reveals that only two of the three wings are needed, Malmstrom AFB becomes the most vulnerable, as it is the lone single mission base of the three.

Our nation's survival guarantees personal freedom and security. While the global war on terror is a day-to-day priority, decision-makers must never take the nation's existence for granted. The triad has enhanced national survival from its outset, a mission as important today as it was during the Cold War.

Air Force Doctrine and Development Center
Air Force Research Institute

AU Quick Look 08-04

Appendix E

Interview Questions

- What will the world look like in 15 years, and what will be the security challenges to the United States in diplomatic, informational, military, and economic terms?
- Given this future conception, what do you expect US national security strategy will look like?
- What enduring qualities make the USAF important to the nation?
- Given your concept of a future US national security strategy and thoughts on USAF enduring qualities, what is your vision for the ideal USAF in 15 years?
- To achieve this vision, what roles, missions, and functions should USAF take on?
- What roles, missions, and functions should the USAF shed?
- What roles, missions, and functions should be retained but accomplished through leveraging the Total Force, to include civilians and/or contractors?
- What will be the most essential role, mission or function of the USAF in 15 years?
- Given our discussion of this future USAF, what key elements would a service strategy to achieve this vision contain?

Appendix F

Personnel Interviewed

Interviews were conducted with three-, two-, and one-star flag officers serving in the Secretariat, Air Staff, AU, AFMC, ACC, AFSPC, NSA, AFR, ANG, HAF, NATO, and French Air Force. Additionally, interviews included selected retired four- and three-star flag officers, military contractors, DOD civilians at the research labs, and "think-tanks," and selected faculty at civilian universities, by rank. These included

Four-star general retired: 1
Three-star generals, retired: 6
Active duty three-star generals: 9
Active duty two-star generals: 4
Active duty one-star generals: 6
Senior Executive Service civilian leaders: 4
Chief executive officers/think-tank subject matter experts: 7

Further, interviews included action officers from HQ USAF, HQ AFSPC, HQ ACC, Eighth and Ninth Air Force, HQ Air Support Operations Center, Joint Special Operations University, USAF Special Operations School, ACC/Coalition and Irregular Warfare Center, State Department, United States Agency for International Development, Treasury Department, Operation Iraqi Freedom (OIF) CAOC, OIF Coalition Air Force Training Team, FBI, Drug Enforcement Agency, AWC, and ASCS faculty and students, plus international officers *via* electronic poll.

Appendix G

A Timeline of USAF Strategic Thought